METHODS IN MOLECULAR BIOLOGY

Series Editor
John M. Walker
School of Life and Medical Sciences
University of Hertfordshire
Hatfield, Hertfordshire, UK

For further volumes:
http://www.springer.com/series/7651

For over 35 years, biological scientists have come to rely on the research protocols and methodologies in the critically acclaimed *Methods in Molecular Biology* series. The series was the first to introduce the step-by-step protocols approach that has become the standard in all biomedical protocol publishing. Each protocol is provided in readily-reproducible step-by-step fashion, opening with an introductory overview, a list of the materials and reagents needed to complete the experiment, and followed by a detailed procedure that is supported with a helpful notes section offering tips and tricks of the trade as well as troubleshooting advice. These hallmark features were introduced by series editor Dr. John Walker and constitute the key ingredient in each and every volume of the *Methods in Molecular Biology* series. Tested and trusted, comprehensive and reliable, all protocols from the series are indexed in PubMed.

3D Bioprinting

Principles and Protocols

Edited by

Jeremy M. Crook

ARC Centre of Excellence for Electromaterials Science, Intelligent Polymer Research Institute, AIIM Facility, Innovation Campus, University of Wollongong, Squires Way, Fairy Meadow, NSW, Australia; Illawarra Health and Medical Research Institute, University of Wollongong, Wollongong, NSW, Australia; Department of Surgery, St Vincent's Hospital, The University of Melbourne, Fitzroy, VIC, Australia

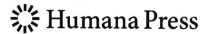 Humana Press

Editor
Jeremy M. Crook
ARC Centre of Excellence
for Electromaterials Science, Intelligent
Polymer Research Institute, AIIM
Facility, Innovation Campus
University of Wollongong
Squires Way, Fairy Meadow, NSW, Australia

Illawarra Health and Medical Research Institute
University of Wollongong
Wollongong, NSW, Australia

Department of Surgery
St Vincent's Hospital, The University of Melbourne
Fitzroy, VIC, Australia

ISSN 1064-3745   ISSN 1940-6029   (electronic)
Methods in Molecular Biology
ISBN 978-1-0716-0522-6   ISBN 978-1-0716-0520-2   (eBook)
https://doi.org/10.1007/978-1-0716-0520-2

This Humana imprint is published by the registered company Springer Science+Business Media, LLC, part of Springer Nature.
The registered company address is: 1 New York Plaza, New York, NY 10004, U.S.A.

Preface

3D bioprinting is a form of additive manufacturing that incorporates a bio-ink comprising cells and/or biocompatible materials for 3D tissue engineering. Using computer-aided design (CAD) for 3D modelling, potentially based on scans taken directly from a patient, bio-inks are printed layer by layer to form scaffolds of desired size and shape, with cells being preserved inside the scaffolds for functional integration, maturation, and tissue formation. Strategies for bioprinting include co-printing of biomaterials with cells for encapsulated cell constructs or printing biomaterial scaffolds that are seeded in vitro or in vivo with cells after printing. Regardless of the approach taken, 3D bioprinting is ideally performed within a regulatory framework of good laboratory practice (GLP) for standardized, optimized, and controlled tissue fabrication, or more rigorous good manufacturing practice (GMP) for clinical-product development. Among other benefits, creating quality "tissues" is the most immediate way to circumvent the limitations of conventional 2D cell culture and human tissue and organ donation for research and therapeutics.

In recognizing the potential of 3D bioprinting, academic and commercial research and development (R&D) groups around the world are investing significantly in bioprinting infrastructure to prepare for the next major phase in researching and commercializing biomaterials and cell-based products for tissue engineering. Importantly, bioprinting need not entail setting up large and expensive facilities, but can involve smaller initiatives to support the activities of entrepreneurial start-ups, individual universities, research institutes, or laboratories. Whatever the scale or purpose, a printing facility should align with global "best practice" for working with biomaterials and cells, safeguarding quality fabrication and augmenting research and translational application. For example, the succession of commercial and clinical aspirations could be facilitated by having a low-cost quality-controlled GLP laboratory for research that can support more expensive clinically compliant GMP activities. In addition, research and clinical-grade variants of a printing process will provide consistency between laboratory and clinical activities for more predictable and better translational outcomes.

This volume brings together contributions from experts in 3D bioprinting, and in turn champions, and facilitates the use of bioprinting for quality R&D and translation. The book is divided into two parts with the first covering generic themes in bioprinting to introduce readers new to the field whilst bolstering understanding of those with experience. The second part includes a collection of complete and standardized protocols for preparing, characterizing, and printing a variety of biomaterials/cells/tissues, with priority given to methods for printing defined and humanized constructs suitable for human tissue modelling and clinical practice.

As a volume in the highly successful *Methods in Molecular Biology* series, it aims to contribute to the development of competence in the subject. By providing information that is necessary to establish 3D bioprinting for research and translation, we hope to encourage the use of this rapidly evolving technology and recommend the volume as a valuable resource for bioprinting laboratories/facilities and those new to the field.

Fairy Meadow, NSW, Australia   *Jeremy M. Crook*

Contents

Contributors

ANTHONY ATALA • *Wake Forest Institute for Regenerative Medicine, Wake Forest University, Winston-Salem, NC, USA*

CHRISTOPH BISIG • *Adolphe Merkle Institute, University of Fribourg, Fribourg, Switzerland; Comprehensive Molecular Analytics, Helmholtz Center Munich, Munich, Germany*

ROMANE BLANCHARD • *BioFab3D@ACMD, St Vincent's Hospital Melbourne, Fitzroy, VIC, Australia; Department of Surgery, St Vincent's Hospital, University of Melbourne, Fitzroy, VIC, Australia*

JUSTIN BOURKE • *@BioFab3D Facility, St Vincent's Hospital Melbourne, Melbourne, VIC, Australia; ARC Centre of Excellence for Electromaterials Science, Intelligent Polymer Research Institute, University of Wollongong, Wollongong, NSW, Australia; Clinical Neurosciences, St. Vincent's Hospital Melbourne, Melbourne, VIC, Australia; Department of Medicine, St. Vincent's Hospital Melbourne, University of Melbourne, Melbourne, VIC, Australia*

PETER CHOONG • *Department of Surgery, St Vincent's Hospital Melbourne, The University of Melbourne, Melbourne, VIC, Australia; @BioFab3D Facility, St Vincent's Hospital Melbourne, Melbourne, VIC, Australia; ARC Centre of Excellence for Electromaterials Science, Intelligent Polymer Research Institute, University of Wollongong, Wollongong, NSW, Australia*

JEREMY M. CROOK • *ARC Centre of Excellence for Electromaterials Science, Intelligent Polymer Research Institute, AIIM Facility, Innovation Campus, University of Wollongong, Squires Way, Fairy Meadow, NSW, Australia; Illawarra Health and Medical Research Institute, University of Wollongong, Wollongong, NSW, Australia; Department of Surgery, St. Vincent's Hospital, The University of Melbourne, Fitzroy, VIC, Australia*

GONZALO DE ARANDA IZUZQUIZA • *Department of Bioengineering and Aerospace Engineering, Universidad Carlos III de Madrid (UC3M), Madrid, Spain*

RAPHAEL DEVILLARD • *Tissue Bioengineering, University of Bordeaux, Bordeaux, France; Tissue Bioengineering, Inserm, Bordeaux, France; Services d'Odontologie et de Santé Buccale, CHU Bordeaux, Bordeaux, France*

CLAUDIA DI BELLA • *BioFab3D@ACMD, St Vincent's Hospital, Melbourne, VIC, Australia; Department of Surgery, St Vincent's Hospital, University of Melbourne, Melbourne, VIC, Australia*

SUSAN DODDS • *ARC Centre of Excellence for Electromaterials Science, Office of the Deputy Vice-Chancellor (Research and Industry Engagement), La Trobe University, Melbourne, VIC, Australia*

SERENA DUCHI • *BioFab3D@ACMD, St Vincent's Hospital Melbourne, Fitzroy, VIC, Australia; Department of Surgery, St Vincent's Hospital, University of Melbourne, Fitzroy, VIC, Australia*

NATHALIE DUSSERRE • *Tissue Bioengineering, University of Bordeaux, Bordeaux, France; Tissue Bioengineering, Inserm, Bordeaux, France*

MANUELA ESTERMANN • *Adolphe Merkle Institute, University of Fribourg, Fribourg, Switzerland*

CORMAC D. FAY • *SMART Infrastructure Facility, Faculty of Engineering and Information Sciences, University of Wollongong, Wollongong, NSW, Australia*

João N. Ferreira • *Exocrine Gland Biology and Regeneration Research Group, Faculty of Dentistry, Chulalongkorn University, Bangkok, Thailand; Faculty of Dentistry, National University of Singapore, Singapore, Singapore*

Jean-Christophe Fricain • *Tissue Bioengineering, University of Bordeaux, Bordeaux, France; Tissue Bioengineering, Inserm, Bordeaux, France; Services d'Odontologie et de Santé Buccale, CHU Bordeaux, Bordeaux, France*

Marta García • *Department of Bioengineering and Aerospace Engineering, Universidad Carlos III de Madrid (UC3M), Madrid, Spain; Division of Epithelial Biomedicine, CIEMAT-CIBERER, Madrid, Spain; Department of Basic Research, Instituto de Investigación Sanitaria de la Fundación Jiménez Díaz, Madrid, Spain*

Erik Gatenholm • *CELLINK LLC, Cambridge, MA, USA*

Eliza Goddard • *ARC Centre of Excellence for Electromaterials Science, Humanities and Social Sciences, La Trobe University, Melbourne, VIC, Australia*

Davit Hakobyan • *Tissue Bioengineering, University of Bordeaux, Bordeaux, France; Tissue Bioengineering, Inserm, Bordeaux, France*

Narutoshi Hibino • *Division of Cardiac Surgery, The Johns Hopkins Hospital, Baltimore, MD, USA; Section of Cardiac Surgery, Department of Surgery, The University of Chicago, Advocate Children's Hospital, Chicago, IL, USA*

Robert M. I. Kapsa • *@BioFab3D Facility, St Vincent's Hospital Melbourne, Melbourne, VIC, Australia; ARC Centre of Excellence for Electromaterials Science, Intelligent Polymer Research Institute, University of Wollongong, Wollongong, NSW, Australia; Clinical Neurosciences, St. Vincent's Hospital Melbourne, Melbourne, VIC, Australia; Department of Medicine, St Vincent's Hospital Melbourne, University of Melbourne, Melbourne, VIC, Australia*

Dongxu Ke • *Wake Forest Institute for Regenerative Medicine, Wake Forest University, Winston-Salem, NC, USA*

Carlos Kengla • *Wake Forest Institute for Regenerative Medicine, Wake Forest University, Winston-Salem, NC, USA*

Olivia Kerouredan • *Tissue Bioengineering, University of Bordeaux, Bordeaux, France; Tissue Bioengineering, Inserm, Bordeaux, France; Services d'Odontologie et de Santé Buccale, CHU Bordeaux, Bordeaux, France*

Keekyoung Kim • *School of Engineering, University of British Columbia, Kelowna, BC, Canada; Department of Mechanical and Manufacturing Engineering, Schulich School of Engineering, University of Calgary, Calgary, AB, Canada*

Magdalena Kita • *@BioFab3D Facility, St Vincent's Hospital Melbourne, Melbourne, VIC, Australia; ARC Centre of Excellence for Electromaterials Science, Intelligent Polymer Research Institute, University of Wollongong, Wollongong, NSW, Australia; Clinical Neurosciences, St. Vincent's Hospital Melbourne, Melbourne, VIC, Australia*

Hitendra Kumar • *School of Engineering, University of British Columbia, Kelowna, BC, Canada*

Verónica López • *Division of Epithelial Biomedicine, CIEMAT-CIBERER, Madrid, Spain*

Hector Martinez • *CELLINK LLC, Cambridge, MA, USA*

Chantal Medina • *Tissue Bioengineering, University of Bordeaux, Bordeaux, France; Tissue Bioengineering, Inserm, Bordeaux, France*

Anushree Mohandas • *BioFab3D@ACMD, St Vincent's Hospital Melbourne, Fitzroy, VIC, Australia*

Andrés Montero • *Department of Bioengineering and Aerospace Engineering, Universidad Carlos III de Madrid (UC3M), Madrid, Spain*

SEAN V. MURPHY • *Wake Forest Institute for Regenerative Medicine, Wake Forest University, Winston-Salem, NC, USA*

CATHERINE NGAN • *Department of Surgery, St Vincent's Hospital Melbourne, The University of Melbourne, Melbourne, VIC, Australia; @BioFab3D Facility, St Vincent's Hospital Melbourne, Melbourne, VIC, Australia; ARC Centre of Excellence for Electromaterials Science, Intelligent Polymer Research Institute, University of Wollongong, Wollongong, NSW, Australia*

CATHAL O'CONNELL • *BioFab3D@ACMD, St Vincent's Hospital Melbourne, Fitzroy, VIC, Australia; ARC Centre of Excellence for Electromaterials Science, Intelligent Polymer Research Institute, University of Wollongong, Wollongong, NSW, Australia; Department of Medicine, St Vincent's Hospital Melbourne, University of Melbourne, Melbourne, VIC, Australia*

HUGO OLIVEIRA • *Tissue Bioengineering, University of Bordeaux, Bordeaux, France; Tissue Bioengineering, Inserm, Bordeaux, France*

CHIN SIANG ONG • *Division of Cardiac Surgery, The Johns Hopkins Hospital, Baltimore, MD, USA; Division of Cardiology, The Johns Hopkins Hospital, Baltimore, MD, USA*

CARMINE ONOFRILLO • *BioFab3D@ACMD, St Vincent's Hospital Melbourne, Fitzroy, VIC, Australia; Department of Surgery, St Vincent's Hospital, University of Melbourne, Fitzroy, VIC, Australia*

IBRAHIM T. OZBOLAT • *Engineering Science and Mechanics Department, Penn State University, University Park, PA, USA; The Huck Institutes of the Life Sciences, Penn State University, University Park, PA, USA; Biomedical Engineering Department, Penn State University, University Park, PA, USA; Materials Research Institute, Penn State University, University Park, PA, USA*

ALKE PETRI-FINK • *Adolphe Merkle Institute, University of Fribourg, Fribourg, Switzerland*

ISAREE PITAKTONG • *Department of Biomedical Engineering, The Johns Hopkins University, Baltimore, MD, USA*

LEON POPE • *BioFab3D@ACMD, St Vincent's Hospital Melbourne, Fitzroy, VIC, Australia*

ANITA QUIGLEY • *@BioFab3D Facility, St Vincent's Hospital Melbourne, Melbourne, VIC, Australia; ARC Centre of Excellence for Electromaterials Science, Intelligent Polymer Research Institute, University of Wollongong, Wollongong, NSW, Australia; Clinical Neurosciences, St. Vincent's Hospital Melbourne, Melbourne, VIC, Australia; Department of Medicine, St Vincent's Hospital Melbourne, University of Melbourne, Melbourne, VIC, Australia*

CRISTINA QUÍLEZ • *Department of Bioengineering and Aerospace Engineering, Universidad Carlos III de Madrid (UC3M), Madrid, Spain*

MURIELLE REMY • *Tissue Bioengineering, University of Bordeaux, Bordeaux, France; Tissue Bioengineering, Inserm, Bordeaux, France*

JUNXIANG REN • *BioFab3D@ACMD, St Vincent's Hospital Melbourne, Fitzroy, VIC, Australia*

BARBARA ROTHEN-RUTISHAUSER • *Adolphe Merkle Institute, University of Fribourg, Fribourg, Switzerland*

SASITORN RUNGARUNLERT • *Department of Preclinical and Applied Animal Science, Faculty of Veterinary Science, Mahidol University, Nakhon Pathom, Thailand*

DEDY SEPTIADI • *Adolphe Merkle Institute, University of Fribourg, Fribourg, Switzerland*

PATRICK THAYER • *CELLINK LLC, Cambridge, MA, USA*

EVA TOMASKOVIC-CROOK • *ARC Centre of Excellence for Electromaterials Science, Intelligent Polymer Research Institute, AIIM Facility, University of Wollongong,*

Wollongong, NSW, Australia; Illawarra Health and Medical Research Institute, University of Wollongong, Wollongong, NSW, Australia

GANOKON URKASEMSIN • Department of Preclinical and Applied Animal Science, Faculty of Veterinary Science, Mahidol University, Nakhon Pathom, Thailand

LETICIA VALENCIA • Department of Bioengineering and Aerospace Engineering, Universidad Carlos III de Madrid (UC3M), Madrid, Spain

DIEGO VELASCO • Department of Bioengineering and Aerospace Engineering, Universidad Carlos III de Madrid (UC3M), Madrid, Spain

GORDON G. WALLACE • @BioFab3D Facility, St Vincent's Hospital Melbourne, Melbourne, VIC, Australia; ARC Centre of Excellence for Electromaterials Science, Intelligent Polymer Research Institute, University of Wollongong, Wollongong, NSW, Australia

KELSEY WILLSON • Wake Forest Institute for Regenerative Medicine, Wake Forest University, Winston-Salem, NC, USA

YANG WU • Engineering Science and Mechanics Department, Penn State University, University Park, PA, USA; The Huck Institutes of the Life Sciences, Penn State University, University Park, PA, USA; School of Mechanical Engineering and Automation, Harbin Institute of Technology (Shenzhen), Shenzhen, China

YIN YU • Institute for Synthetic Biology, Shenzhen Institutes of Advanced Technology, Chinese Academy of Sciences, Shenzhen, People's Republic of China; University of Chinese Academy of Sciences, Beijing, People's Republic of China

YAHUI ZHANG • Adhezion Biomedical, LLC, Wyomissing, PA, USA

YIFAN ZHANG • BioFab3D@ACMD, St Vincent's Hospital Melbourne, Fitzroy, VIC, Australia

Part I

Generic Themes in 3D Bioprinting

Chapter 1

History and Trends of 3D Bioprinting

Patrick Thayer, Hector Martinez, and Erik Gatenholm

Abstract

The field of bioprinting is rapidly evolving as researchers innovate and drive the field forward. This chapter provides a brief overview of the history of bioprinting from the first described printer system in the early 2000s to present-day relatively inexpensive commercially available units and considers the current state of the field and emerging trends, including selected applications and techniques.

Key words 3D bioprinting, 4D bioprinting, Cell printing, Tissue engineering, Biomaterials, Bioinks

1 Introduction

Long in the realm of science fiction, the fabrication of tissues and organs is quickly becoming a reality since the turn of the century. The practical origin of bioprinting can be traced to Thomas Boland's group at Clemson University with the development of the first bioprinter in the early 2000s. This achievement was made possible through the modification of a commercially available inkjet printer to deposit cells as opposed to ink. This system could pattern a bioink consisting of a mixture of cells, culture medium, and serum into a cell culture dish [1, 2]. Further improvements in both the system and the bioink composition allowed for the deposition of cells within more viscous biological hydrogels to permit the fabrication of structures within a third dimension [3]. Similarly at the University of Missouri, Gabor Forgacs developed another bioprinting approach through the positioning of individual cell spheroids in patterns [4, 5]. After positioning the spheroids, they began to fuse together during culture, resulting in the assembly of thicker tissues. Furthermore, the first 3D bioprinting company was formed in 2007 from this lab, known as Organovo, which sought to develop and commercialize tissue models for drug screening and disease modelling. This approach is similar to another technology commercialized by Cyfuse of Japan in 2011. Cyfuse assembled 3D structures through spheroids supported by a needle array known

Jeremy M. Crook (ed.), *3D Bioprinting: Principles and Protocols*, Methods in Molecular Biology, vol. 2140,
https://doi.org/10.1007/978-1-0716-0520-2_1, © Springer Science+Business Media, LLC, part of Springer Nature 2020

as the Kenzen method [6]. In 2007 RegenHu was founded in Switzerland and focused on extrusion based 3D bioprinter systems known as the Biofactory. Similarly, Envision-Tec which traditionally manufactures traditional 3D printers developed its 3D-Bioplotter system. While EnvisionTec and RegenHu develop advanced bioprinter systems that are sold to researchers that allow them to develop their own tissue constructs. Alternatively, other companies have commercialized other bioprinting methods such as laser-assisted deposition as commercialized by Poietis who focuses on the engineering of skin tissue mimics [7].

While bioprinter systems quickly became commercially available following their initial development, due to the high price of what was essentially a robotic dispensing system, their uptake was largely limited to commercial entities such as pharmaceutical companies with academic researchers generally unable to afford available systems. Despite the eventual wider application, from the outset the risk of what was essentially an unsubstantiated technology was too high for adoption by industry. The researchers that could validate the technology were priced out, further hindering acceptance and application. Coinciding with the birth of the "maker" movement that provided cheap and easy to use 3D printers, researchers began to build their own in the lab. The first cost-effective bioprinters were variations of open source systems or commercially available MakerBots or Ultimaker systems. Researchers began to eliminate the demand for quarter-million dollar machines by demonstrating that replacement of the heated filament head of a traditional 3D printer with a syringe pump or pneumatic extrusion system can obtain similar results. This democratization of bioprinting research nurtured a new industry of low-cost, affordable bioprinters that were developed and refined by the researchers and field as a whole rather than a small "moneyed" group. Concurrently, new research areas were developed around the printable biomaterials, known as bioinks. The most predominant companies in this area include CELLINK originating in Boston, MA, USA, Allevi (formerly BioBots) from Philadelphia, PA, USA, and Se3D from the Bay Area in California. CELLINK provides dual- and triple-headed bioprinter systems, the INKREDIBLE series and BIOX series, respectively. The BIOX series in particular contains interchangeable print-heads that allows the use of diverse bioprinting modalities such as pneumatic extrusion (heated and cooled), thermoplastic printing, mechanical extrusion, and inkjet printing. Additionally, the systems include clean chamber technology that allows bioprinting in a clean environment on the benchtop without the need for a biosafety cabinet. Allevi produces bioprinter systems most notabily the Allevi 2 which contains two heated pneumatic print-heads. While SE3D produces systems geared toward the high school education market with the R3bel mini and with some printers such as the R3bel X aimed specifically toward the research market.

Additionally, several companies have focused on developing the bioinks for use with the 3D bioprinter systems. This is part of a broader effort to begin to standardize materials for bioprinting but also the bioprinting protocols to enable the development of true bioprinting applications. Most predominant examples of these companies include CELLINK, Advanced Biomatrix, and Allevi. Both CELLINK and Allevi pair bioprinter systems with bioinks while Advanced Biomatrix focuses on biomaterial development solely. The development of bioinks is key for the advancement of the bioprinting field as a whole. Both easy to print, and consistent compositions are necessary for the myriad and tissues that can be generated through bioprinting. They will play a key role in any printing protocol and its reproducibility.

2   Transitioning from 2D to 3D Cell Culture

Two dimensional (2D) cell culture techniques have enabled many great discoveries related to cell biology, genetics, cancer, and cell differentiation. Despite the success in basic science, shortfalls exist in the utilization of 2D cell culture techniques to evaluate drugs and therapies for commercial and clinical applications [8]. 2D cell culture techniques lack the complexity of actual tissues and organ systems. While animal models have been employed to test drugs in a preclinical setting, there has been a desire to move away from and phase out animal modeling due to high costs, ethical concerns, and poor translation to clinical applications [9]. Presently, there is a platform that sits in between 2D cell culture and animal modeling and which can address the limitations of earlier techniques [8, 10] by culturing cells in a three-dimensional environment. By comparison, cells in 2D cultures exhibit nonphysiological morphology compared to in vivo and 3D cell culture resulting in different mechanics [11, 12]. Additionally, cells in 3D exhibit different signaling cascades and sensitivities to different drugs [11, 12]. The failure of traditional technologies can be attributed to the simplicity of 2D cell culture compared to the actual in vivo environment. In particular, gradients in nutrients, oxygen, and other factors found in native tissue are largely nonexistent, eliminating many stimuli that guide cellular behavior [13]. In recent years, extensive efforts have been directed towards the development of 3D cell culture techniques, underpinned by bioprinting as a technique for enabling the fabrication of complex, multifaceted constructs that can begin to resemble the immensely complex native tissue environment [14, 15]. Secondly, bioprinting can be used to facilitate the automation of the fabrication of tissue constructs for basic research, tissue engineering, therapeutic, or drug screening applications. Techniques such as bioprinting will enable the shifting of 3D tissue construct fabrication from a manual process to a fully automated workflow; similar to other fields such as cellular assayology involving machine handling for faster and more efficient cell screening. Similarly, bioprinting will facilitate the automated fabrication of reproducible multifaceted constructs to be used in research and, in the longer-term, translation in conjunction with 3D cell culture techniques. Importantly, bioprinting systems possess the precision necessary to incorporate patterning of matrix, cells, and bioactive cues within these 3D constructs, permitting the engineering of multifaceted templates that can support the generation of tissues.

3   Revolutionizing Tissue Engineering

Tissue engineering over the past several decades has focused around determining viable combinations of scaffolding biomaterials, cell populations, and the necessary biological, mechanical, and chemical factors that can guide the maturation of an engineered tissue [16, 17]. However, continued research in the tissue engineering field has exposed limitations of this tissue engineering modality. It has become apparent that the engineering of suitable tissue constructs has proven to be more difficult than initially believed. Several limitations that have emerged can be traced to shortfalls in existing biofabrication technologies and an inability to fabricate more complicated, heterogeneous structures reproducibility. The emergence of new techniques such as bioprinting has the potential to address several of these challenges in tissue construct fabrication. The immediate areas where bioprinting can address the existing challenges in tissue engineering are outlined below.

3.1   Internal Architecture of Soft Constructs

In traditional fabrication approaches, biomaterials are pipetted into well plates as droplets or cast into molds, resulting nonorganized internal architectures. In order to impart organization, postfabrication manipulation must be performed, including gelation under flow or applied shear [18], application of magnetic fields [19], or incorporation of microfiber networks [20, 21]. In contrast, bioprinting can be utilized to incorporate internal organization and architecture into tissue constructs. At a basic level, this internal organization can be integrated through the modulation of the infill parameters when a digital model to a printable G-code file. These most common infills include patterns such as grid or rectilinear, concentric, or hexagonal configurations which at a basic level can roughly model different tissue architectures (Fig. 1). Furthermore, they can influence the underlying mechanical characteristics of the resulting construct by modulating the orientation, density, and arrangement of the individual filaments [22, 23]. The resulting

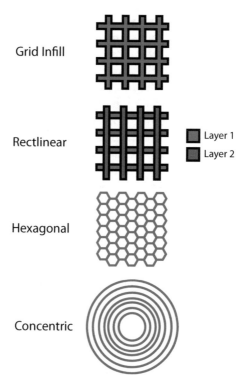

Grid Infill

Rectlinear

Layer 1
Layer 2

Hexagonal

Concentric

Fig. 1 Common infill patterns of soft constructs including grid, rectilinear, hexagonal, and concentric configurations

filament characteristics (diameter, uniformity, and spacing) can be further modulated by the translation speed, the applied pressure, and nozzle diameter during the printing process. Most notably, the orientation of the filaments that make up the infill pattern has a predominant effect on cells seeded within or on the surface [24, 25]. This imparted topography both within the construct and on its surface can guide cell alignment, organization, and ultimately differentiation [24–26].

Furthermore, bioprinting can be utilized to fabricate constructs that contain two or more distinct cell types, matrices, or other materials in controlled specified arrangements such as alternating layers or individual alternating filaments or droplets. Using bioprinting, the spacing between the individual filaments and the thickness of the individual layers can be controlled. In particular, increased spacing can be utilized to introduce porosity and void regions that can facilitate tissue in growth or nutrient influx. This basic spatial control provides the groundwork for the development of more specialized internal architectures that better mimic native tissue microstructures such as neuromuscular junctions [27], kidney nephron units [28], blood vessels [29], and heart valves [29]. Ultimately, these can be utilized for tissue specific applications or study how cells communicate with each other [30–32] or drive their migration [33, 34], proliferation, or differentiation [35].

3.2 Vascular Networks

Most tissues in the body contain an extensive vascular network to provide nutrients and oxygen to the cells within the tissues. Recapitulating vascularization during the biofabrication process is critical for the engineering of robust functional tissue constructs. One of the most predominant challenges in tissue construct fabrication is the difficulty in incorporating a vascular network that can sufficiently provide nutrients deep within the tissue. As a rule of thumb, diffusion is most effective in providing nutrients and oxygen to cells that are within 200 μm of a vessel wall or construct boundary. Beyond this distance, the inclusion of void regions or hollow conduits becomes necessary to support cell viability. Bioprinting can enable the fabrication of tissue constructs that contain these void regions or conduits through several approaches (Fig. 2). Void regions can be incorporated into constructs via control over spacing of individual filaments as done with the bioprinting of lattice structures. Most commonly, these can be generated through a rectilinear, grid, or hexagonal infill patterning to create different void region geometries. Another approach utilizes sacrificial bioinks for the incorporation of conduit or channel networks within the construct bulk [36–38]. Bioinks such as pluronics can be patterned within the construct to act as perfusion channels within the constructs to enable deeper penetration of nutrients and oxygen. Furthermore, these channel networks can serve as a template to support vascularization via endothelial cell seeding and induced angiogenesis. One such approach to induce angiogenesis is deposition of a morphogen [39] (such as VEGF) containing bioinks in proximity to these channels that can induce angiogenesis. The incorporation of perfusable networks and factors can be difficult through traditional cell culture and fabrication approaches. Bioprinting will allow researchers better control over network fabrication and incorporation and begin to address the vascularization challenge.

3.3 Heterogeneous Constructs and Gradients

No tissue found in the body consists of homogenous compositions of its extracellular matrix, cell phenotype, and mechanical characteristics. Tissues contain these gradients that are critical in the physiological functionality of native tissues and the broader organizational structure of the tissue. Importance of the ECM organization can be found in its role cell migration, proliferation, and differentiation. For example, gradients have a predominant role in guiding wound healing [40, 41] and tissue development [42, 43]. From a tissue engineering perspective, regional-differences in cell types, matrix organization and composition, and mechanical properties are found in nearly every organ. Bioprinter systems can fabricate constructs consisting of region-specific depositions of biomaterials, cells, and growth factors. This enables the patterning of physiologically relevant gradients into these tissue constructs for a wide range of applications, including the following:

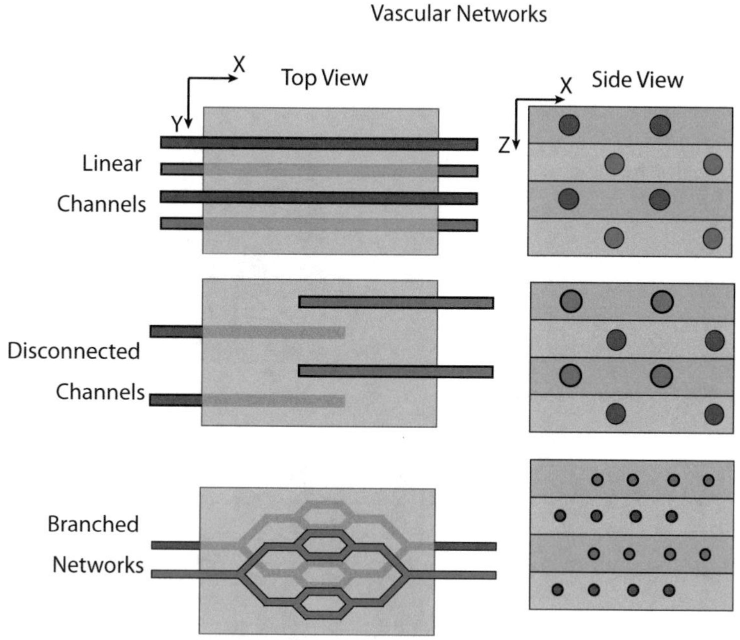

Fig. 2 Bioprinting can enable the fabrication of tissue constructs with void regions or conduits that form linear or disconnected channels or more complex branched networks

- Guidance of cell migration and proliferation [44, 45].
- Induction of differentiation into targeted phenotypes in specific regions [46–48].
- Engineering of specialized constructs that serve to integrate adjacent tissue types [49, 50].

Heterogeneous constructs fabricated through bioprinting can be utilized on a more basic level to study cells in three dimensions (Fig. 3). This ability to pattern any or all of a cellular, extracellular matrix, or biochemical stimuli component can be utilized in applications such as the following:

- Cell–cell cross talk between adjacent or distant droplets or filaments.
- Cell migration driven by growth factor or matrix gradients.
- Gradient formation.
- Matrix composition on cell behavior and cross talk between other cells.
- Organoid or spheroid formation.
- Multilayered constructs for study of stratification.

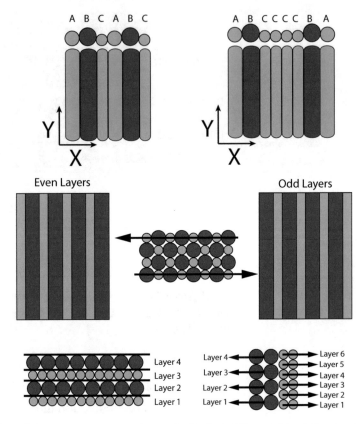

Fig. 3 Heterogeneous constructs can be fabricated through bioprinting to study cells in 3D

4   Bioinks—A New Class of Biomaterials

Biomaterials that are suitable for bioprinting are known as biological inks or bioinks [51, 52]. Bioinks can be considered a distinct subclass of biomaterials due to their ability to be patterned in 3D without an external support. Traditional biomaterials have been developed to be pipetted manually into well plates or molds. These low viscosity biomaterials are not suitable for use in a bioprinter system due to do a lack of structural rigidity prior to crosslinking. In order to build 3D dimensional constructs, novel materials must be developed, or existing biomaterials must be adapted that can fulfill these roles. Adapting traditional biomaterials to achieve what is known as printability can be a challenge. A suitable bioink must have several characteristics beyond enabling cell viability such as the following:

- Can be extruded in a controlled fashion to generate cylindrical shapes or rigid droplets.

- Exhibit shear-thinning characteristics to flow under an applied force but retain shape under no loading.

- Ability to cross-link or self-assemble to prevent dissolution and maintain shape when multilayered or stacked.

These additional characteristics distinguish bioinks from traditional biomaterials. In recent years, bioinks have been primarily derived from traditional biomaterials such as [53], but not limited to, alginates, gelatins, collagens, polyethylene glycols, and silk. In the formulation of these traditional biomaterials as bioinks, several trends have emerged. One of the most popular and successful approaches has been the use of thickeners to increase the viscosity of these biomaterials and impart shear thinning characteristics [54, 55]. These thickeners can include additives such as nanocellulose, high molecular weight biomolecules such as polyethylene oxide and gelatin or inorganic additives such as clays. Other popular approaches include the thermoregulation of these bioinks. Gelatin is the textbook example of this concept, it is necessary to maintain the temperature of a gelatin based bioink within its thermal transition region to permit extrusion and rapid solidification upon deposition. If the bioink is too warm it will extrude as a liquid and not retain its shape after printing. Conversely, if it is too cold, it will extrude unevenly and at high pressures. Furthermore, the adjacent filaments will not bind together, compromising the stability of the resultant structure. A final approach has been bioprinting into a shear thinning support bath rather than changing the characteristics of the bioink. This approach, known as the FRESH method, utilizes slurries to support a lower viscosity bioink after deposition [51].

Cross-linking modalities have focused around several key concepts (Fig. 4):

- Cross-linking by addition of an ionic solution such as calcium. This process is often reversible but may require special considerations both during and after the bioprinting process.

- Cross-linking through the application of a light source whether in the UV or visible range. This process is often nonreversible due to the use of a catalyst that chemically binds biomolecules together. This approach is the most popular approach due to its rapid action and flexibility in application.

- Cross-linking through enzymatic activity is an approach that often takes advantage of natural biological processes. An example of this approach is the assembly of fibrin gels.

- Cross-linking through self-assembly is another approach for imparting stability to a bioink. This approach is often governed

Pre-Crosslinking   Post-Crosslinking

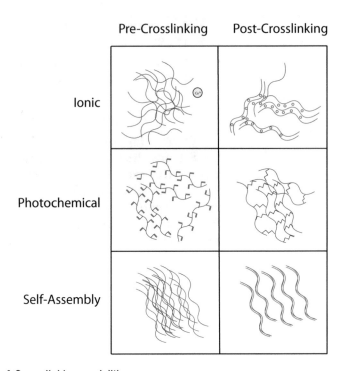

Ionic

Photochemical

Self-Assembly

Fig. 4 Cross-linking modalities

by environmental factors such as temperature, PH, and biomolecule concentration. Examples of this approach include thermosensitive assembly of gelatin and self-assembly of collagen fibers.

5   Emerging Fields of Bioprinting

Beyond mainstream research in bioprinting, there are several other emerging fields worth considering that relate to material development and applications.

5.1   4D Bioprinting

Researchers have begun to develop bioinks that are more dynamic after bioprinting. These bioinks are a specialized class of biomaterials known as 4D biomaterials which change their characteristics with time or stimuli [56, 57]. Most commonly, these 4D materials can fold or change shape in response to environmental changes such as PH, light, or osmotic pressure. This is relevant for tissue engineering as a whole because it allows scaffolding materials to dynamically stimulate the incorporated cells without the need for an external mechanism. Furthermore, it can be utilized for have a temporal and controlled change in stiffness, growth factor release, or degradation during tissue development. From a bioprinting perspective it allows the fabrication of simpler or flat structures that fold into the final confirmation. This allows for more efficient bioprinting as it removes the effect of gravity during the printing process and may allow the utilization of biomaterials that otherwise could not be printable. While early in their conceptualization and application, 4D bioinks have the potential to disrupt the field and be applied for a variety of target tissue applications [58, 59].

5.2 Soft Robotics

Soft robotics is another emerging field for bioprinting [60, 61]. Printable 4D biomaterials discussed in the previous section can find immediate application in soft robotics as mechanically robust backbone structures for a soft robotics development. Moreover, bioprinting can fabricate tissue constructs that contain internal architectures that can be contracted. Similar to 4D biomaterials that can contract via external stimulations, bioinks containing cells that are mechanically active serve the same purpose. For example, constructs could comprise patterned oriented filaments prepared via control over the infill characteristics and containing myocytes in an aligned or crimped structure. The maturation of the cells in combination with the templated internal architecture could allow the development of a more physiological muscle tissue. This can be combined with bioprinted PDMS or other silicones that may contain conductive components or neurogenic bioinks to allow neuronal integration. The maturation of such a construct could result in a biological robot that contracts when stimulated by the neuron or repeatability. These structures could be utilized as artificial muscle or for other applications. While this is one idea, one can imagine that bioprinting can be utilized to build other types of bio robots and soft robots [62].

5.3 Organ-on-a-Chip

As discussed previously, 3D bioprinting enables the repeatable fabrication of multidimensional scaffolds and tissue constructs. These constructs can be utilized as small-scale models for various tissues and organs in the body. This allows researchers to study cells more in their physiological environment due to the ability to dimensionally pattern the cells to better mimic tissue structure in situ. 3D bioprinting has an immediate application in the Organ-on-a-Chip field [63]. The technique can enhance currently utilized organ-on-a-chip systems and enable new research based on the technology. Briefly, the aim of organ-on-a-chip is to connect miniaturized tissues or organs through a microchannel network [28, 64, 65]. An organ-on-a-chip simulates entire organs or systems in order to study their activity, response, and/or mechanics in isolation or conjugation with other organs. For example, this analytical tool allows researchers to test drugs or other therapies on cells derived from the patient within a system that can mimic native organ systems. This allows one to evaluate systematic effects of drugs on several tissue types or study the interactions between different organs or cell types more readily [66]. Bioprinting will play a key role in this field. First and foremost, bioprinting can fabricate physiological-mimetic miniaturized organs and organ systems via the incorporation of microarchitecture by patterning of cells and matrix. Furthermore, standardization of the organ-on-a-chip design for specific applications would permit the use of bioprinting as a fabrication technique for these tools. One could envision a diagnostic laboratory of the future where technicians bioprint patient-specific miniaturized organ systems to test therapies prior to their prescription. These models could be utilized for all sorts of therapies such as oncology, wound healing, and chronic diseases to both better treat patients and generate new therapies.

5.4 Vascularized Tissues

One of the greatest promises with bioprinting is the ability to generate and incorporate vascular networks within a tissue construct [37, 67]. The lack of a vascular networks to provide nutrients to cells within a tissue engineered construct and remove waste has been a major limitation in the traditional tissue engineering field. The development of sacrificial bioinks that can both generate channels but also provide the necessary cues to induce the formation of a vascular phenotype is necessary for the advancement of the field. Future directions in this area will include the integration of a vascularized construct with an external bioreactor system to drive the flow of nutrients. Additionally, it may be necessary to generate a capillary bed or large vessels that can integrate with surrounding tissue at implantation or the surgeon can surgically attach.

6 Future 3D Culture Infrastructure Driven by Bioprinting

One can imagine a future where 3D bioprinting will be routinely applied for tissue models, organ on a chip systems, dynamic biomaterials, and advanced organoid generation are in biomedical research. However, in addition to bioprinting, other technological advances will be necessary to accommodate shifting from 2D to 3D cell-based R&D platforms including advanced 3D microscopy, genetic and protein assays, culture systems such as bioreactors, and noninvasive monitoring systems. Adaption of these other technologies will enable bioprinting to be applied to its full potential as a biomanufacturing technique to fabricate patterned constructs that serve as templates for tissue development, drug screening, and other advanced therapies such as personalized medicine.

7 Conclusions

The field of bioprinting is rapidly changing and evolving as the technology is placed in the hands of more and more researchers that can drive the field forward. While this introduction considers the current state of the field and emerging trends, it is far from complete. Subsequent chapters and sections of this volume will further detail specific applications and techniques.

References

1. Mironov V, Boland T, Trusk T, Forgacs G, Markwald RR (2003) Organ printing: computer-aided jet-based 3D tissue engineering. Trends Biotechnol 21(4):157–161. https://doi.org/10.1016/S0167-7799(03)00033-7

2. Roth EA, Xu T, Das M, Gregory C, Hickman JJ, Boland T (2004) Inkjet printing for high-throughput cell patterning. Biomaterials 25 (17):3707–3715. https://doi.org/10.1016/j.biomaterials.2003.10.052

3. Xu T, Jin J, Gregory C, Hickman JJ, Boland T (2005) Inkjet printing of viable mammalian cells. Biomaterials 26(1):93–99. https://doi.org/10.1016/j.biomaterials.2004.04.011

4. Mironov V, Visconti RP, Kasyanov V, Forgacs G, Drake CJ, Markwald RR (2009) Organ printing: tissue spheroids as building blocks. Biomaterials 30(12):2164–2174. https://doi.org/10.1016/j.biomaterials.2008.12.084

5. Norotte C, Marga FS, Niklason LE, Forgacs G (2009) Scaffold-free vascular tissue engineering using bioprinting. Biomaterials 30 (30):5910–5917. https://doi.org/10.1016/j.biomaterials.2009.06.034

6. Moldovan NI, Hibino N, Nakayama K (2017) Principles of the Kenzan method for robotic cell spheroid-based three-dimensional bioprinting. Tissue Eng Part B Rev 23 (3):237–244. https://doi.org/10.1089/ten.TEB.2016.0322

7. Pages E, Remy M, Keriquiel V, Correa M, Guillotin B, Guillemot F (2015) Creation of highly defined mesenchymal stem cell patterns in three dimensions by laser-assisted bioprinting. J Nanotechnol Eng Med 6(2):21006

8. Breslin S, O'Driscoll L (2016) The relevance of using 3D cell cultures, in addition to 2D monolayer cultures, when evaluating breast cancer drug sensitivity and resistance. Oncotarget 7(29):45745–45756. https://doi.org/10.18632/oncotarget.9935

9. Begley CG, Ellis LM (2012) Drug development: raise standards for preclinical cancer research. Nature 483(7391):531–533. https://doi.org/10.1038/483531a

10. Breslin S, O'Driscoll L (2013) Three-dimensional cell culture: the missing link in drug discovery. Drug Discov Today 18 (5–6):240–249. https://doi.org/10.1016/j.drudis.2012.10.003

11. Sun T, Jackson S, Haycock JW, MacNeil S (2006) Culture of skin cells in 3D rather than 2D improves their ability to survive exposure to cytotoxic agents. J Biotechnol 122 (3):372–381. https://doi.org/10.1016/j.jbiotec.2005.12.021

12. Yamada KM, Cukierman E (2007) Modeling tissue morphogenesis and cancer in 3D. Cell 130(4):601–610. https://doi.org/10.1016/j.cell.2007.08.006

13. Griffith LG, Swartz MA (2006) Capturing complex 3D tissue physiology in vitro. Nat Rev Mol Cell Biol 7(3):211–224. https://doi.org/10.1038/nrm1858

14. Kang HW, Lee SJ, Ko IK, Kengla C, Yoo JJ, Atala A (2016) A 3D bioprinting system to produce human-scale tissue constructs with structural integrity. Nat Biotechnol 34 (3):312–319. https://doi.org/10.1038/nbt.3413

15. Murphy SV, Atala A (2014) 3D bioprinting of tissues and organs. Nat Biotechnol 32 (8):773–785. https://doi.org/10.1038/nbt.2958

16. Bartold PM, McCulloch CA, Narayanan AS, Pitaru S (2000) Tissue engineering: a new paradigm for periodontal regeneration based on molecular and cell biology. Periodontol 24:253–269

17. Yang S, Leong KF, Du Z, Chua CK (2001) The design of scaffolds for use in tissue engineering. Part I. Traditional factors. Tissue Eng 7(6):679–689. https://doi.org/10.1089/107632701753337645

18. Aubin H, Nichol JW, Hutson CB, Bae H, Sieminski AL, Cropek DM, Akhyari P, Khademhosseini A (2010) Directed 3D cell alignment and elongation in microengineered hydrogels. Biomaterials 31(27):6941–6951. https://doi.org/10.1016/j.biomaterials.2010.05.056

19. Guo C, Kaufman LJ (2007) Flow and magnetic field induced collagen alignment. Biomaterials 28(6):1105–1114. https://doi.org/10.1016/j.biomaterials.2006.10.010

20. Butcher AL, Offeddu GS, Oyen ML (2014) Nanofibrous hydrogel composites as mechanically robust tissue engineering scaffolds. Trends Biotechnol 32(11):564–570. https://doi.org/10.1016/j.tibtech.2014.09.001

21. Thayer PS, Verbridge SS, Dahlgren LA, Kakar S, Guelcher SA, Goldstein AS (2016) Fiber/collagen composites for ligament tissue engineering: influence of elastic moduli of sparse aligned fibers on mesenchymal stem cells. J Biomed Mater Res A 104 (8):1894–1901. https://doi.org/10.1002/jbm.a.35716

22. Fernandez-Vicente M, Calle W, Ferrandiz S, Conejero A (2016) Effect of infill parameters on tensile mechanical behavior in desktop 3D printing. 3D Print Addit Manufact 3 (3):183–192

23. Lubombo C, Huneault M (2018) Effect of infill patterns on the mechanical performance of lightweight 3D-printed cellular PLA parts. Mater Today Commun 17:214–228

24. Maiullari F, Costantini M, Milan M, Pace V, Chirivi M, Maiullari S, Rainer A, Baci D, Marei HE, Seliktar D, Gargioli C, Bearzi C, Rizzi R (2018) A multi-cellular 3D bioprinting approach for vascularized heart tissue engineering based on HUVECs and iPSC-derived cardiomyocytes. Sci Rep 8(1):13532. https://doi.org/10.1038/s41598-018-31848-x

25. Tijore A, Irvine SA, Sarig U, Mhaisalkar P, Baisane V, Venkatraman S (2018) Contact guidance for cardiac tissue engineering using 3D bioprinted gelatin patterned hydrogel. Biofabrication 10(2):025003. https://doi.org/10.1088/1758-5090/aaa15d

26. Tsukamoto Y, Akagi T, Shima F, Akashi M (2017) Fabrication of orientation-controlled 3D tissues using a layer-by-layer technique and 3D printed a thermoresponsive gel frame. Tissue Eng Part C Methods 23(6):357–366. https://doi.org/10.1089/ten.TEC.2017.0134

27. Karande TS, Ong JL, Agrawal CM (2004) Diffusion in musculoskeletal tissue engineering scaffolds: design issues related to porosity, permeability, architecture, and nutrient mixing. Ann Biomed Eng 32(12):1728–1743. https://doi.org/10.1007/s10439-004-7825-2

28. Homan KA, Kolesky DB, Skylar-Scott MA, Herrmann J, Obuobi H, Moisan A, Lewis JA (2016) Bioprinting of 3D convoluted renal proximal tubules on Perfusable chips. Sci Rep 6:34845. https://doi.org/10.1038/srep34845

29. Liu H, Zhou H, Lan H, Liu T, Liu X, Yu H (2017) 3D printing of artificial blood vessel: study on multi-parameter optimization design for vascular molding effect in alginate and gelatin. Micromachines (Basel) 8(8):237. https://doi.org/10.3390/mi8080237

30. Byron A, Randles MJ, Humphries JD, Mironov A, Hamidi H, Harris S, Mathieson PW, Saleem MA, Satchell SC, Zent R, Humphries MJ, Lennon R (2014) Glomerular cell cross-talk influences composition and assembly of extracellular matrix. J Am Soc Nephrol 25 (5):953–966. https://doi.org/10.1681/ASN.2013070795

31. Coulouarn C, Corlu A, Glaise D, Guenon I, Thorgeirsson SS, Clement B (2012) Hepatocyte-stellate cell cross-talk in the liver engenders a permissive inflammatory microenvironment that drives progression in hepatocellular carcinoma. Cancer Res 72 (10):2533–2542. https://doi.org/10.1158/0008-5472.CAN-11-3317

32. Jobling P, Pundavela J, Oliveira SM, Roselli S, Walker MM, Hondermarck H (2015) Nerve-Cancer cell cross-talk: a novel promoter of tumor progression. Cancer Res 75 (9):1777–1781. https://doi.org/10.1158/0008-5472.CAN-14-3180

33. Bourget JM, Kerouredan O, Medina M, Remy M, Thebaud NB, Bareille R, Chassande O, Amedee J, Catros S, Devillard R (2016) Patterning of endothelial cells and mesenchymal stem cells by laser-assisted bioprinting to study cell migration. Biomed Res Int 2016:3569843. https://doi.org/10.1155/2016/3569843

34. Huang TQ, Qu X, Liu J, Chen S (2014) 3D printing of biomimetic microstructures for cancer cell migration. Biomed Microdevices 16(1):127–132. https://doi.org/10.1007/s10544-013-9812-6

35. Byambaa B, Annabi N, Yue K, Trujillo-de Santiago G, Alvarez MM, Jia W, Kazemzadeh-Narbat M, Shin SR, Tamayol A, Khademhosseini A (2017) Bioprinted osteogenic and vasculogenic patterns for engineering 3D bone tissue. Adv Healthc Mater 6(16). https://doi.org/10.1002/adhm.201700015

36. Jia W, Gungor-Ozkerim PS, Zhang YS, Yue K, Zhu K, Liu W, Pi Q, Byambaa B, Dokmeci MR, Shin SR, Khademhosseini A (2016) Direct 3D bioprinting of perfusable vascular constructs using a blend bioink. Biomaterials 106:58–68. https://doi.org/10.1016/j.biomaterials.2016.07.038

37. Lee VK, Kim DY, Ngo H, Lee Y, Seo L, Yoo SS, Vincent PA, Dai G (2014) Creating perfused functional vascular channels using 3D bio-printing technology. Biomaterials 35 (28):8092–8102. https://doi.org/10.1016/j.biomaterials.2014.05.083

38. Lee VK, Lanzi AM, Haygan N, Yoo SS, Vincent PA, Dai G (2014) Generation of multi-scale vascular network system within 3D hydrogel using 3D bio-printing technology. Cell Mol Bioeng 7(3):460–472. https://doi.org/10.1007/s12195-014-0340-0

39. Cui H, Zhu W, Nowicki M, Zhou X, Khademhosseini A, Zhang LG (2016) Hierarchical fabrication of engineered vascularized bone biphasic constructs via dual 3D bioprinting: integrating regional bioactive factors into

architectural design. Adv Healthc Mater 5 (17):2174–2181. https://doi.org/10.1002/adhm.201600505

40. Arnold F, West DC (1991) Angiogenesis in wound healing. Pharmacol Ther 52 (3):407–422. https://doi.org/10.1016/0163-7258(91)90034-j

41. Steed DL (1997) The role of growth factors in wound healing. Surg Clin North Am 77 (3):575–586. https://doi.org/10.1016/s0039-6109(05)70569-7

42. Bier E, De Robertis EM (2015) Embryo development. BMP gradients: a paradigm for morphogen-mediated developmental patterning. Science 348(6242):aaa5838. https://doi.org/10.1126/science.aaa5838

43. Naba A, Clauser KR, Ding H, Whittaker CA, Carr SA, Hynes RO (2016) The extracellular matrix: tools and insights for the "omics" era. Matrix Biol 49:10–24. https://doi.org/10.1016/j.matbio.2015.06.003

44. Devreotes P, Horwitz AR (2015) Signaling networks that regulate cell migration. Cold Spring Harb Perspect Biol 7(8):a005959. https://doi.org/10.1101/cshperspect.a005959

45. Haeger A, Wolf K, Zegers MM, Friedl P (2015) Collective cell migration: guidance principles and hierarchies. Trends Cell Biol 25 (9):556–566. https://doi.org/10.1016/j.tcb.2015.06.003

46. Faia-Torres AB, Guimond-Lischer S, Rottmar M, Charnley M, Goren T, Maniura-Weber K, Spencer ND, Reis RL, Textor M, Neves NM (2014) Differential regulation of osteogenic differentiation of stem cells on surface roughness gradients. Biomaterials 35 (33):9023–9032. https://doi.org/10.1016/j.biomaterials.2014.07.015

47. Wang L, Li Y, Huang G, Zhang X, Pingguan-Murphy B, Gao B, Lu TJ, Xu F (2016) Hydrogel-based methods for engineering cellular microenvironment with spatiotemporal gradients. Crit Rev Biotechnol 36 (3):553–565. https://doi.org/10.3109/07388551.2014.993588

48. Wang PY, Clements LR, Thissen H, Tsai WB, Voelcker NH (2015) Screening rat mesenchymal stem cell attachment and differentiation on surface chemistries using plasma polymer gradients. Acta Biomater 11:58–67. https://doi.org/10.1016/j.actbio.2014.09.027

49. Han F, Zhou F, Yang X, Zhao J, Zhao Y, Yuan X (2015) A pilot study of conically graded chitosan-gelatin hydrogel/PLGA scaffold with dual-delivery of TGF-beta1 and BMP-2 for regeneration of cartilage-bone interface. J Biomed Mater Res B Appl Biomater 103 (7):1344–1353. https://doi.org/10.1002/jbm.b.33314

50. Samavedi S, Vaidya P, Gaddam P, Whittington AR, Goldstein AS (2014) Electrospun meshes possessing region-wise differences in fiber orientation, diameter, chemistry and mechanical properties for engineering bone-ligament-bone tissues. Biotechnol Bioeng 111 (12):2549–2559. https://doi.org/10.1002/bit.25299

51. Groll J, Burdick JA, Cho DW, Derby B, Gelinsky M, Heilshorn SC, Jungst T, Malda J, Mironov VA, Nakayama K, Ovsianikov A, Sun W, Takeuchi S, Yoo JJ, Woodfield TBF (2018) A definition of bioinks and their distinction from biomaterial inks. Biofabrication 11(1):013001. https://doi.org/10.1088/1758-5090/aaec52

52. Williams D, Thayer P, Martinez H, Gatenholm E, Khademhosseini A (2018) A perspective on the physical, mechanical and biological specifications of bioinks and the development of functional tissues in 3D bioprinting. Bioprinting 9:19–36

53. Hospodiuk M, Dey M, Sosnoski D, Ozbolat IT (2017) The bioink: a comprehensive review on bioprintable materials. Biotechnol Adv 35 (2):217–239. https://doi.org/10.1016/j.biotechadv.2016.12.006

54. Markstedt K, Mantas A, Tournier I, Martinez Avila H, Hagg D, Gatenholm P (2015) 3D bioprinting human chondrocytes with Nanocellulose-alginate bioink for cartilage tissue engineering applications. Biomacromolecules 16(5):1489–1496. https://doi.org/10.1021/acs.biomac.5b00188

55. Holzl K, Lin S, Tytgat L, Van Vlierberghe S, Gu L, Ovsianikov A (2016) Bioink properties before, during and after 3D bioprinting. Biofabrication 8(3):032002. https://doi.org/10.1088/1758-5090/8/3/032002

56. Hilderbrand AM, Ovadia EM, Rehmann MS, Kharkar PM, Guo C, Kloxin AM (2016) Biomaterials for 4D stem cell culture. Curr Opin Solid State Mater Sci 20(4):212–224. https://doi.org/10.1016/j.cossms.2016.03.002

57. Miao S, Cui H, Nowicki M, Lee SJ, Almeida J, Zhou X, Zhu W, Yao X, Masood F, Plesniak MW, Mohiuddin M, Zhang LG (2018) Photolithographic-stereolithographic-tandem fabrication of 4D smart scaffolds for improved stem cell cardiomyogenic differentiation. Biofabrication 10(3):035007. https://doi.org/10.1088/1758-5090/aabe0b

58. Castro NJ, Meinert C, Levett P, Hutmacher D (2017) Current developments in

multifunctional smart materials for 3D/4D bioprinting. Curr Opin Biomed Eng 2:67–75

59. Ong CS, Nam L, Ong K, Krishnan A, Huang CY, Fukunishi T, Hibino N (2018) 3D and 4D bioprinting of the myocardium: current approaches, challenges, and future prospects. Biomed Res Int 2018:6497242. https://doi.org/10.1155/2018/6497242

60. Kim S, Laschi C, Trimmer B (2013) Soft robotics: a bioinspired evolution in robotics. Trends Biotechnol 31(5):287–294. https://doi.org/10.1016/j.tibtech.2013.03.002

61. Majiki C (2014) Soft robotics: a perspective—current trends and prospects for the future. Soft Robot 1(1):5–11

62. Wehner M, Truby RL, Fitzgerald DJ, Mosadegh B, Whitesides GM, Lewis JA, Wood RJ (2016) An integrated design and fabrication strategy for entirely soft, autonomous robots. Nature 536(7617):451–455. https://doi.org/10.1038/nature19100

63. Lee H, Cho DW (2016) One-step fabrication of an organ-on-a-chip with spatial heterogeneity using a 3D bioprinting technology. Lab Chip 16(14):2618–2625. https://doi.org/10.1039/c6lc00450d

64. Knowlton S, Yenilmez B, Tasoglu S (2016) Towards single-step biofabrication of organs on a Chip via 3D printing. Trends Biotechnol 34(9):685–688. https://doi.org/10.1016/j.tibtech.2016.06.005

65. Zhang YS, Arneri A, Bersini S, Shin SR, Zhu K, Goli-Malekabadi Z, Aleman J, Colosi C, Busignani F, Dell'Erba V, Bishop C, Shupe T, Demarchi D, Moretti M, Rasponi M, Dokmeci MR, Atala A, Khademhosseini A (2016) Bioprinting 3D microfibrous scaffolds for engineering endothelialized myocardium and heart-on-a-chip. Biomaterials 110:45–59. https://doi.org/10.1016/j.biomaterials.2016.09.003

66. Bhise NS, Manoharan V, Massa S, Tamayol A, Ghaderi M, Miscuglio M, Lang Q, Shrike Zhang Y, Shin SR, Calzone G, Annabi N, Shupe TD, Bishop CE, Atala A, Dokmeci MR, Khademhosseini A (2016) A liver-on-a-chip platform with bioprinted hepatic spheroids. Biofabrication 8(1):014101. https://doi.org/10.1088/1758-5090/8/1/014101

67. Kolesky DB, Homan KA, Skylar-Scott MA, Lewis JA (2016) Three-dimensional bioprinting of thick vascularized tissues. Proc Natl Acad Sci U S A 113(12):3179–3184. https://doi.org/10.1073/pnas.1521342113

Chapter 2

Cell Processing for 3D Bioprinting: Quality Requirements for Quality Assurance in Fundamental Research and Translation

Jeremy M. Crook

Abstract

Bioprinting is an additive manufacturing process where biomaterials-based inks are printed layer-by-layer to create three-dimensional (3D) structures that mimic natural tissues. Quality assurance for 3D bioprinting is paramount to undertaking fundamental research and preclinical and clinical product development. It forms part of quality management and is vital to reproducible and safe tissue fabrication, function, and regulatory approval for translational application. This chapter seeks to place the implementation of quality practices in 3D bioprinting front-of-mind, with emphasis on cell processing, although important to all components and procedures of the printing pipeline.

Key words 3D bioprinting, Cells, Quality assurance, Quality management system, Good laboratory practice, Good manufacturing practice, Clinical compliance

1 Introduction

The synergism of cell biology (particularly stem cell biology) and biomaterial technology for 3D bioprinting promises to have a profound impact on cell-based research and translation for tissue engineering and regenerative medicine. Bioprinting is rapidly advancing underpinned by novel printable materials that display properties which in a precise and physiological fashion can facilitate cell survival, state, and fate before, during, and after printing. Thus, by combining cells with materials to 3D-print tissue systems, we aim to recapitulate the in vivo microphysiological environment of immature and mature tissues, inclusive of molecular events involved in the cellular production, interaction and clearance of molecules within healthy and pathologic conditions, being representative of the cell niche (e.g., the inner cell mass (ICM) of a blastocyst for pluripotent stem cells, walls of the lateral ventricles and subgranular zone of the dentate gyrus of the hippocampus for neural stem cells, renal capsule of the kidney for renal stem cells, or vascular niche in bone marrow for hematopoietic stem cells). These events translate

Jeremy M. Crook (ed.), *3D Bioprinting: Principles and Protocols*, Methods in Molecular Biology, vol. 2140,
https://doi.org/10.1007/978-1-0716-0520-2_2, © Springer Science+Business Media, LLC, part of Springer Nature 2020

to influencing in situ cell behavior and functionality by altering morphology, adhesion, motility, proliferation, endocytotic activity, protein abundance, and gene regulation. At a most basic level, printed biomaterials can be used to recapitulate the role of extracellular matrix (ECM) whereby the printed material will transmit specific signals to cells that are decoded into biochemical signals. Key properties of materials include surface properties such as topography, which can influence on a micro- and nanometer scale, mechanical properties such as material modulus or softness, morphological properties including porosity and shape, and electrical properties, which effect via material actuation or intracellular ion (e.g., calcium) influx [1–3]. Not surprisingly, the complex process of cell and material interfacing for printing can be difficult to optimize and reproduce and so requires a clear understanding and control of both biological and nonbiological components, with the potential for variability in cell survival and function especially significant. Optimised and quality assured (QA) cell processing for bioprinting is therefore critical in both research and translational product development, requiring quality controlled (QC) systems and procedures from initial cell procurement and holding, through to cell culture, preprinting cell integration with biomaterials for bioink preparation, or postprint cell seeding to printed scaffolds, differentiation, and in-process characterization. My own work has involved both encapsulation and post-print cell seeding approaches for generating human stem cell based tissues, with examples including bioink development with human neural stem cell and induced pluripotent stem cell (iPSC) encapsulation for extrusion printing presently detailed in Chapters 10 and 17 [1, 2], as well as iPSC seeding onto printed GelMA for development of human brain organoids [4]. Regardless of the approach, though, processing cells intended for use in bioprinting toward quality research and clinical product development critically requires an established QA procedure providing a validated, formalized, and regulated methodology, designed to afford adequate confidence that the entire operation will fulfil expected and defined requirements.

This chapter aims to draw attention to the quality requirements for the cellular component of bioprinting, albeit it as important for all other aspects of bioprinting in both fundamental research and preclinical and clinical product development. By doing so, it seeks to place the implementation of quality practices front of mind, earlier rather than later, although ideally from the outset of project planning and management, to actual printed tissue prototyping and clinical product development. Ultimately, the goal is to facilitate progress, instilling confidence amongst current and future adopters of bioprinting technology as well as other stakeholders.

2 Quality Assurance and Management 101

QA through QC for cell processing is a way to mitigate the risk of making mistakes during the procurement, handling, and culture of cells used for bioprinting, which ultimately can result in a defective tissue product. Both are integral to a quality management plan and monitored by a quality management system (QMS). A QMS should be implemented that documents the organizational structure, responsibilities, policies, procedures, processes and resources required for QA. The QMS should be based on the principles of current good laboratory practice (cGLP) or consummate manufacturing practice (cGMP), and should consider relevant local regulatory requirements and guidance. However, such systems are not necessarily required to be performed under a cGLP/cGMP license, but should at least meet a standard that ensures fit-for-purpose and reproducibility of the cell processing including storage (be it cryostorage or otherwise), and critically establishes traceability for all components and procedures used from the point of informed consent for procurement of primary tissues or cells, to establishment of cell stocks, to bioprinted tissue employed for preclinical research and development (R&D) through to commercial or clinical product development. Therefore, all critical procedures used in delivery of the cells for bioprinting and beyond should be documented as formally recorded standard operating procedures (SOPs), associated forms, and higher level documents such as policies, process descriptions covering a number of SOPs, manuals and training documents. All records should be controlled to assure that only the correct and current procedures and documentation are used and that old versions are archived carefully to allow review and audits in the future.

3 Processing for Quality Assured iPSCs and Bioprinting

If we consider more closely iPSCs as an example of a versatile cell type able to be used for bioprinting, their applications beyond research include 3D cell systems for drug discovery and tissue replacement therapy, with the latter extending to autologous or allogeneic application (Fig. 1).

Figure 2 shows a processing map for QA iPSCs, adapted from previously published cGMP clinical-grade human embryonic stem cell derivation and banking [5], highlighting the myriad of steps required for deriving cell lines, with several critical break-points and numerous points where formal checks can and should be made for risk mitigation. Each point in the process requires one or more SOPs for completion, with SOPs also required for testing. In some instances testing can be as basic as bright-field microscopy for

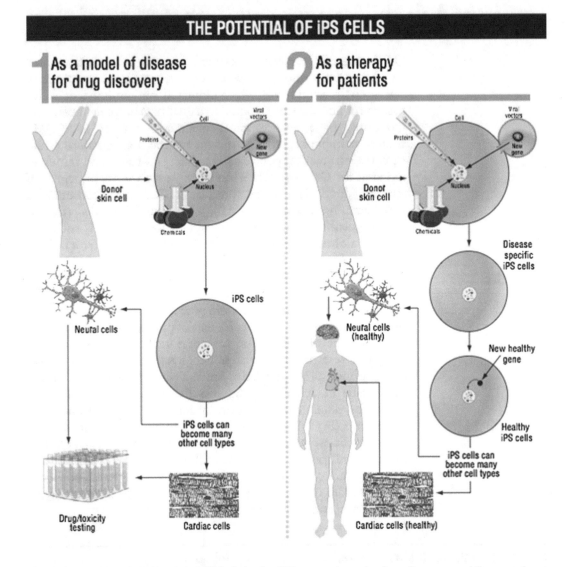

Fig. 1 Applications of 3D bioprinted iPSC-derived cell/tissue systems for drug discovery and tissue replacement (autologous or allogeneic) therapy

observation of iPSC colony morphology or more complex assayology such as immunochemistry-based analyses, karyotyping chromosome anomalies, and endotoxin testing.

Figure 3 shows a processing map for bioprinting cells, which follows on from iPSC processing depicted in Fig. 2, and includes printing stem cells that are subsequently differentiated in situ as well as printing predifferentiated stem cells. Among other things, it is important to emphasize sterility throughout processing, extending to both cells and biomaterials, with the risk of introducing contaminants at virtually any point from start to finish.

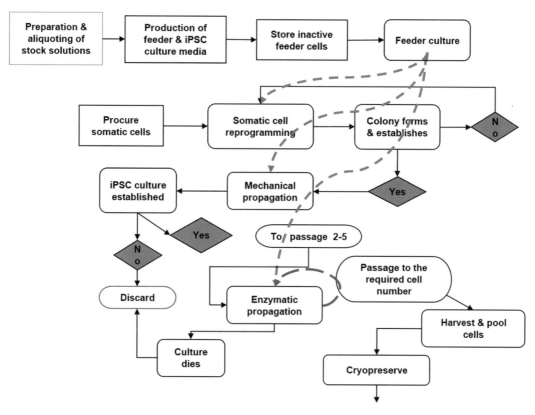

Fig. 2 Processing map for QA iPSC line derivation, culture, and banking

4   Processing for Bioprinted Cell and Tissue Therapy

Placed in the context of a cell therapy product supply chain, relatively complex clinical-grade cell and tissue processing comprises a series of dynamic elements starting in a clinical facility (e.g., operating room), subsequent cGMP bioprocessing and printing, and then returning to the clinic (Fig. 4). Patients or donors must first be screened for health and for safety (identification) and once cleared, biopsy collection occurs with a proper inventory made for tracking purposes. Biopsies would be transported to a manufacturing facility for cell extraction, with isolated cells being initially put into storage (short-term and/or long-term, depending on whether they are meant for immediate use and/or banking). Again, using the example of an iPSC-based therapy, somatic cell isolation would be followed by expansion toward reprogramming for iPSC line derivation, with working stem cell samples subsequently used for tissue fabrication by, for example, direct-write printing and differentiation of a printed construct. Importantly, the entire process should employ cGMP-grade reagents and

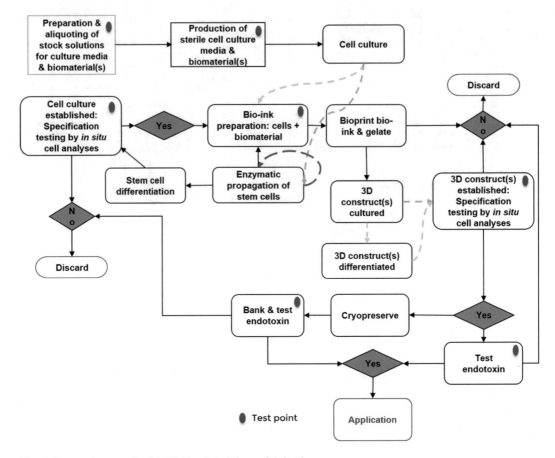

Fig. 3 Processing map for QA 3D bioprinted tissue fabrication

consumables for clinical-compliance. Once processing is complete and a tissue is produced with the right specifications, it would be moved to the clinical facility, for administration to the patient recipient.

5   Cost Caveat and Conclusion

It is important to be aware of the financial cost involved when implementing any QA system for bioprocessing, with the highest not surprisingly associated with a cGMP-grade process. However, it is worth considering the significant longer-term savings (both in terms of time and money) made by building a level of quality into a processing platform early on (be it for research or other) so as to avoid having to reinvent the wheel when something goes wrong because of a failure to meet a certain standard or specification. Of course, more than just a minimum level of QA/QC is ostensibly the order of the day with at least implementation of GLPs for cell processing and bioprinting seemingly justified.

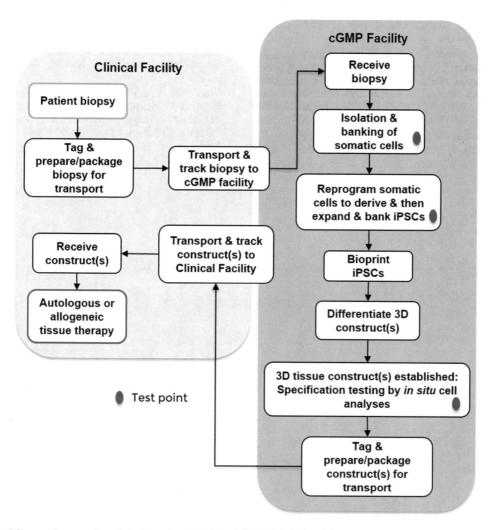

Fig. 4 Processing map for clinical-grade cGMP bioprinted iPSC-derived tissue fabrication, bioprocessing, and therapeutic application

Acknowledgments

The author wishes to acknowledge funding from the Australian Research Council (ARC) Centre of Excellence Scheme (CE140100012).

References

1. Gu Q, Tomaskovic-Crook E, Lozano R, Chen Y, Kapsa RM, Zhou Q et al (2016) Functional 3D neural mini-tissues from printed gel-based human neural stem cells. Adv Healthc Mater 5:1429–1438

2. Gu Q, Tomaskovic-Crook E, Wallace GG, Crook JM (2017) 3D bioprinting human induced pluripotent stem cell constructs for in situ cell proliferation and successive multi-

lineage differentiation. Adv Healthc Mater 6:1700175

3. Tomaskovic-Crook E, Zhang P, Ahtiainen A, Kaisvuo H, Lee CY, Beirne S et al (2019) Human neural tissues from neural stem cells using conductive biogel and printed polymer microelectrode arrays for 3D electrical stimulation. Adv Healthc Mater 8(15):1900425

4. Tomaskovic-Crook E, Crook JM (2018) Clinically-amendable defined and rapid induction of human brain organoids from induced pluripotent stem cells. In: Turksen K (ed) Organoids, Methods in Molecular Biology Series, 1st edn. Springer Nature, New York, p 1576

5. Crook JM, Peura TT, Kravets L, Bosman AG, Buzzard JJ, Horne R et al (2007) The generation of six clinical-grade human embryonic stem cell lines. Cell Stem Cell 1(5):490–494

Chapter 3

Computer-Aided Design and Manufacturing (CAD/CAM) for Bioprinting

Cormac D. Fay

Abstract

Three-dimensional (3D) printing of human tissues and organs has been an exciting area of research for almost three decades [Bonassar and Vacanti. J Cell Biochem. 72(Suppl 30–31):297–303 (1998)]. The primary goal of bioprinting, presently, is achieving printed constructs with the overarching aim toward fully functional tissues and organs. Technology, in hand with the development of bioinks, has been identified as the key to this success. As a result, the place of computer-aided systems (design and manufacturing—CAD/CAM) cannot be underestimated and plays a significant role in this area. Unlike many reviews in this field, this chapter focuses on the technology required for 3D bioprinting from an initial background followed by the exciting area of medical imaging and how it plays a role in bioprinting. Extraction and classification of tissue types from 3D scans is discussed in addition to modeling and simulation capabilities of scanned systems. After that, the necessary area of transferring the 3D model to the printer is explored. The chapter closes with a discussion of the current state-of-the-art and inherent challenges facing the research domain to achieve 3D tissue and organ printing.

Key words Computer-aided Design, Design, Computer-aided Manufacturing, CAD, CAM, 3D, Bioprinting

1 Introduction

Computer-aided design (CAD) is widely recognized as the key to modern design and fabrication processes and is consequently a multibillion dollar industry [1–3]. It lies at the forefront between the user and the target end product, enabling the designer to translate their ideas into virtual representations. This offered a number of advantages over its predecessor including rapid drafting, design optimization, and most notably virtual simulations, which yields insight into the performance/behavior of the product(s) over time within a simulated environment [4]. Once fully assessed, the product is later realized through the use of its well-known counterpart, that is, Computer-aided manufacturing (CAM). The

Jeremy M. Crook (ed.), *3D Bioprinting: Principles and Protocols*, Methods in Molecular Biology, vol. 2140,
https://doi.org/10.1007/978-1-0716-0520-2_3, © Springer Science+Business Media, LLC, part of Springer Nature 2020

combination of CAD and CAM (CAD/CAM) has proven to be disruptive for the fabrication of target products [5].

While many excellent reviews exist within the literature regarding the topic of bioprinting [6–8], attention is often given to the biological rather than the technical aspects of bioprinting. This is understandable as the target audience primarily relates to the former and the fundamental research challenges are perceived to be within the biological domain [9]. However, it must be noted that in order for the field to achieve its long-term goals, a true multidisciplinary effort must take place. This was identified by Murphy and Atala [9] by classifying three main stages in the bioprinting process, which are (a) imaging and design, (b) choice of materials and cells, and (c) printing of the tissue construct.

The intention behind the material within this chapter is to give the reader an overarching view of the development process behind bioprinting. Considering that bioinks and [customized] printers will be discussed in later chapters, this chapter focuses on the means and methods of data generation, processing, refinement, management, and ultimately the generation of code suitable for interpretation by bioprinters, that is, the CAD/CAM elements for successful bioprints. The primary aim here is to present a high level description of the major elements of CAD/CAM for bioprinting. Specifically, its intent is to give researchers—primarily from the biological side—insight into the technological protocols involved. Considering that it is well recognized that the future of the field and its success depends upon multidisciplinary efforts, it is important that researchers gain an understanding of the protocols involved within the CAD/CAM construct. This chapter aims not only to achieve this but also to provide the reader with the current state of the relevant research elements in addition to the major challenges ahead.

This chapter is structured into six sections and follows the trend that the printing of inert materials has established. Figure 1 presents an overview that typically governs this process and applies to the structure of this chapter. The same process is expected to take place with biomaterials with a number of caveats and challenges along the way. The first section sets the context and background of CAD/-CAM, briefly from its origins to the current development of 3D printing technology. Requiring the "blueprints" for printing tissues or organs, a review of the capturing methodologies is introduced followed by denoising and extraction of a digital representation via segmentation approaches. What follows is a trend on simulations of the captured 3D model. Management and preparation of the resulting data for instructions to the printer is subsequently presented. A final word on the grand challenges appearing in this field and the important place that CAD/CAM processes and procedures follows.

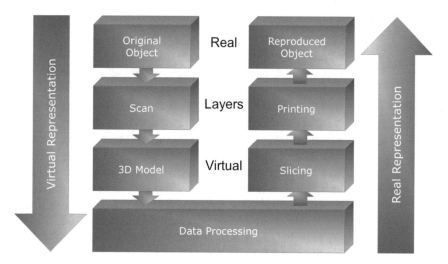

Fig. 1 Overview of the bioprinting process. Building of the virtual representation of an object has corresponding elements when developing the real representation. Both macro processes go through three stages, that is, real, layers (technology), and the virtual space

2 Background

2.1 CAD/CAM

Dr. Patrick J. Hanratty is often credited as the *father of CAM* [10]. Dating back to 1957, he developed a software called PRONTO (Program for Numerical Tooling Operations)—the first commercial Computer Numerical Control (CNC or NC) programming system—during his time at General Electric [11]. The first CAD software, called Sketchpad (A Man-Machine Graphical Communication System), was developed by Dr. Ivan Sutherland— widely credited as the *father of computer graphics*—during his PhD at MIT in 1964 [12, 13]. Its significance paved the first milestone of a user successfully interacting with a computer in order to design objects using a graphical construct.

The impact of Hanratty's and Sutherland's discoveries laid the foundations for what is known today as CAD/CAM technology. At the time computers were prohibitively expensive for consumer audiences. This, coupled with the high engineering demands for vehicular design, the earliest commercial adopter of this technology was the aerospace and automotive industries [14]. The microprocessor industry was next to benefit from this technology [15], which in an almost symbiotic manner, gave rise to increased computational power for further advances of CAD/CAM, thus facilitating the increasing integration of CAD and CAM processes.

This paradigm shift in development approaches has had a tremendous positive impact upon the manufacturing industry [16], with many recognizing the transition as a revolution in modern design and development methods [17, 18]. Reviews in the literature have documented the impact that CAD/CAM has had in various fields such as dentistry [19–21], the social sciences [22], and casting of jewelry [23], among others. One such example of extreme relevance to this discussion is the development of the 3D printer.

2.2 3D Printing

In 1986, Charles Hull applied the concept of papier-mâché to the creation of 3D objects, but in a much more sophisticated manner. His method, which he identified as "stereolithography," consisted of the development of solid 3D structures through the progressive layering of materials "fused" (cured) together using ultraviolet light [24]. This method is now generally referred to as additive manufacturing (AM); distinguishing it from conventional machining methods involving subtractive processes [25].

The idea of AM has expanded into a number of forms including Hull's stereolithography (SLA) along with fused deposition modeling (FDM), selective laser sintering (SLS), selective laser melting (SLM), electronic beam melting (EBM), digital light processing (DLP), laminated object manufacturing (LOM), binder jetting (BJ), and material jetting (MJ) [26, 27]. Overall, the most prevalent model in use is FDM of thermoplastics such as acrylonitrile butadiene styrene (ABS) and polylactic acid (PLA), which is due in no small part to the consumer sector of bench top 3D printers [3]. Reaching this technology readiness level (TRL) was based on the understanding of such materials and the capability of predicting their behavior during the printing processes.

Based on the above technology and the development of more advanced materials suitable for 3D printing, the movement of 3D printing for on-body applications began to emerge, albeit with the use of "inert" materials. For example, reports of bio-related body parts have entered the literature, for example, ears [28], hands [29], sockets for below-knee amputees [30], and prosthetics for swimming [31], among others.

2.3 3D Bioprinting

In 1993, a paper was published in Science by Langer and Vacanti detailing the very first characterization and application of biodegradable 3D scaffolds [32]. This was a breakthrough step for tissue engineering and bioprinting. In the first decade following this publication a number of conventional manufacturing techniques were applied to form 3D scaffolds [33]. However, these approaches did not result in adequate scaffolds required for cell growth, that is, sufficient porosity and a network for their interconnection playing a significant role [34]. The next decade proved to make significant strides in this area—mainly due to the adoption of 3D printing methodologies [35] capable of precise control of scaffolds and their porous structures [36]. Considering that tissues usually possess changes in porosity within their construct, it is of utmost importance to match the mechanical strength and stiffness within this construct for the target tissue structure [37]. CAD naturally plays a significant factor in this design and development in addition to the employed bioinks. While bioinks will be the subject of subsequent chapters, the remainder of this chapter focuses on key elements within the CAD/CAM process for tissue engineering.

3   Medical Imaging

Since the application of X-rays to "look" inside the body in a noninvasive manner, the use of this technology has predominantly been used for medical diagnostic purposes [38]. Since then the medical imaging area has exploded into a rich research discipline of its own right [39]. Its relevance to bioprinting, and CAD thereof, is twofold. The first is the ability to use this technology in order to assess the porosity and internetworking within scaffolds [40]. The second is capturing a 3D model of tissues or organs, which yields necessary information for understanding the interactions of biosystem. As the field of bioprinting expands, the need to reproduce working tissues (and ultimately organs) will require an understanding of the capturing techniques involved in order to transfer a human organ into a 3D representation.

3.1   Computed Tomography (CT)

CT is a process of capturing a large number of tomographic (derived from Ancient Greek τόμος tomos, "slice or section") images of the body using two-dimensional X-ray capturing technology [41]. The differences in absorption by various tissues are represented by a small volume (or voxel) in the data set [42]. This approach is advantageous in terms of its high spatial resolution and short scanning time. However, its use of ionizing radiation yields additional risks for the patient [43].

3.2   Magnetic Resonance Imaging (MRI)

MRI uses nuclear magnetic resonance (NMR) to generate 2D images of the human body (similar to CT scans). The presence of a strong magnetic field can cause certain atomic nuclei to absorb and emit RF energy and align with the applied field [44]. The advantage of this approach over CT is that it does not use ionizing radiation and therefore fewer risks are associated for many target scans.

Functional MRI (fMRI) is a branch of the MRI scan type with its primary application on neurological (brain and spinal cord) activity [45]. Analysis of this type typically monitor changes in blood flow (i.e., the hemodynamic response) by repeatedly measuring the same area multiple even hundreds of times, making the resulting data sets for subsequent processing large and complex. For bioprinting, this approach can yield valuable information related to the dynamics associated with organs and therefore their reproduction.

3.3 Positron Emission Tomography (PET)

PET was the technique of choice to study organ functionality prior to the introduction of fMRI. PET involves the use of radioactive tracers. While it is related to the principle of CT, it differs in the sense that CT is based on transmission, whereas PET is based on emission.

3.4 Ultrasound

Ultrasound has been used for decades in medical imaging and is referred to as medical sonography or ultrasonography [46]. The advantages of this technique over CT or MRI are its affordability [47], portability [48], in addition to its lower risk without the need for ionizing radiation or the use of high power levels [49]. Its place in relation to bioprinting has been reviewed by Zhou [50].

3.5 Other Imaging Platforms

While the above have touched on the major scanning modalities currently related to bioprinting, there are many other techniques available including single photon emission computed tomography (SPECT), diffusion tensor imaging (DTI), optical imaging, or microscopy. A review by Eklund et al. can offer insight into these imaging modalities [51] for further reading.

It is clear that there are a number of options when choosing scanning technology and employed methodology. Overall, the major players in this area are CT and MRI, or derivatives thereof. While this is still a strong area of research, it still has to firmly establish its connection to bioprinting before this process is viable. One considerable challenge is the roll out of technologies for cellular level scans. Moreover, the data processing techniques are yet to be fully established before reaching this level of detail.

4   Noise

It is well accepted that noise reduction (denoising) is a necessary process in medical imaging processing since its outset [52]. This is due, primarily, to noise generated from the instrument itself. For example, because of the ionizing radiation involved during CT scans and the exposure to patients, the power is often reduced to an acceptable health risk level. However, this inherently has an effect on the resulting images requiring the use of processing algorithms to counteract it. Whether this is to aid medical diagnosis via experts in the field or automatic processing and analysis, denoising is a necessary precursor.

A review of the literature will yield the most commonly employed denoising algorithms, which include anisotropic diffusion [53], adaptive filtering [54], bilateral filtering [55], and non-local means [56]. Considering that this is an attractive area of research, one can find studies on improvements to specific scanning modalities, for example, for CT [57], PET [58], MRI [59], and ultrasound [60]. In the context of bioprinting, such noise can have a substantial effect on the segmentation stages and if left unchecked, it can result in misclassifications and/or artifacts in the final print. As a result, additional challenges rests in segmentation procedures, while reliant upon noise reduction yet also be robust enough to account for a certain amount, hence the continued research in this area.

5 Segmentation

Segmentation (or labeling) is the process by which a digital image is partitioned into multiple segments (set of pixels or volume—voxels) in order to produce a representation that is meaningful or easier to analyze [61]. 2D image segmentation has had a long history of research and development [62]. Its origins can arguably be traced back to Roberts in 1963 [63] or Brice in 1970 [64] and is still an active area of research [65]. Its impact to modern society cannot be underestimated and underpins many areas of imaging including satellite imagery [66], vehicle recognition [67], and video compression [68].

With breakthrough segmentation approaches arguably slowing down in the 2D arena, it has been applied to the more challenging area of 3D segmentation for medical imaging. In this arena the goal of segmentation is to use the 2D captured tomographic images and reconstruct them into a 3D model, whereupon an algorithm is applied post or in line to isolate data representing organs, neurological structures (brain or spinal cord), bones, blood vessels, or for tumor detection. This is still an active area of research with no single segmentation algorithm available that can solve all problems. An addition challenge rests with misclassifications. Unlike within the 2D realm or the 3D diagnostic area, which is predominantly for observation and can be omitted or interpolated by the observer, artifacts and imperfections cannot be tolerated in 3D bioprinting due to the formation of unknown obstructions or complications.

Successes in the area of 3D organ/tissue segmentation are vast, and one can find excellent reviews in this area of choice. The reason for this is that researchers have adopted an approach to address one organ at a time and study the inherent challenges therein. For example, Smith [69] and Yazdani [70] examined the literature and yielded excellent reviews on brain segmentation. Petitjean's review is often cited as one of the best for heart segmentation [71] in addition to a more recent review by Slomka [72]. The trends for all segmentation algorithms have moved from standard approaches toward either fully automated segmentation or faster algorithms to reduce computational costs. A similar trend can be seen for kidneys [73], liver [74], or part of the cardiovascular system [75].

A recent trend has moved away from focusing on singular organs to systems capable of segmenting an entire body. An early attempt by Dogdas published an article entitled "Digimouse" demonstrating a capability to capture an entire body, albeit in conjunction with cryosection data [76]. More recently, Lavdas released a paper demonstrating a fully automatic, whole-body, multiorgan segmentation in healthy volunteers [77]. It is also worth noting the National Library of Medicine's (NLM) Visible Human Project, which involves the development of a fully 3D representation of a human male and female from a large library of tomographic data sets [78]. The 3D Heart Library is another great example and is highly recommended. These, and others, have great potential for establishing a basis for bioprinting organs and its awareness is therefore of great importance for future research in this area.

6   Modeling and Simulation

As discussed previously, a major factor of the success of the current consumer based 3D printers is the establishment of how the materials behave (e.g., thermoplastics) and control thereof. As a result, predictions or simulations of the printing process and final outcome can take place on a computer before the printing process commences. The phrase "in silico"" was term coined in 1989 that applies to this very process [79]. More than this, CAM software packages often feature simulation software options to estimate how the CAD model behaves under different environmental conditions over time. The top two approaches are finite element analysis (FEA) [4] and computational fluid dynamics (CFD) [80]. The basic philosophy for both approaches (in addition to others) are the same, that is, the discretization of a large system into smaller pieces that model the entire system and solved with appropriate mathematical formulae. While there can be an overlap with the two methods, generally FEA typically focuses on structural applications, whereas CFD focuses on fluid flow.

Applying the capabilities of CAD/CAM for the capturing, modeling, and simulation of an organ in a virtual manner, before printing, is intuitively an attractive proposition. This could be a powerful approach if applied to bioprinting, and it would reduce the risks associated with the printing of cells and cell-laden structures, giving the user a good estimation/prediction of how biostructures will form within its target environment. Ultimately, when knowledge in the bioformation domain has been firmly established, for example, through biomimicry, autonomous self-assembly, and mini-tissue blocks [9], the basic building blocks for fully functional organ printing is expected to take place. Some examples of this modeling technique have appeared in the literature with important steps taken by Sun with their Bio-CAD modeling package [81], Lacroix examined FEA for bone tissue engineering [82], Chai applied an FE approach for modeling bladders [83], and Noritomi presented a comparison between BioCAD and other contenders [84]. The next horizon is likely to address to the complexities with organ printing, where a combination of structural and fluidic systems is expected. As a result, the line between FEA and CFD in the application domain will merge when tackling complex challenges such as tissue engineering and later organ printing.

For the medical diagnostic area this is known as digital pathology [85, 86]. In the classic instance, this can mean using digital imaging to aid experts in diagnoses. However, for the context of this discussion, the target use is to perform automatic diagnoses and analysis on the captured 3D images/models (phantoms), that is, to (a) investigate whether the virtual organ is compiled correctly, (b) detect whether abnormalities such as tumors exist, and (c) perform predictive analysis on the virtual organ to determine its functionality within the target environment. While this is, or should be, the overarching vision for bioprinting, there has been some research done toward this goal in this area recently.

More recently, examples related to the study of dynamic models from tomographic data are emerging. For instance, Paganelli developed a virtual 4D system (CoMBAT), capable of validating MRI acquired organs with a dynamic model of lungs (digital breathing) as the exemplar [87]. Wang introduced a similar idea to study the motion of the liver [88]. For cardiac motion Veress applied the FEA approach to establish its function [89] and Mukai simulated blood flow and pressure change within the aorta [90]. Cardio/respiratory motion of a thorax phantom was published by Bowlin [91]. Dao studied a multimodal and multiphysical model for estimating patient-specific lumbar support [92]. With other examples are available in the literature, the message is that simulation studies is a young branch of medical diagnosis and the exciting first steps are currently being taken. With progressive stages taking place one must keep in mind that currently the dynamic phantom studies are at a stage of simulating the tissue types for high-level motion analysis, but not tackling the cellular activity or necessary vascularity within organs. This opens up exciting opportunities to virtually construct, analyze, and possibly print organs in the near future.

7 Data Management

Considering the vast adoption of 3D medical capturing technology and its prediction to increase [93], the computational costs associated with data processing, storage, transfer, access, and security is extensive and of concern [94, 95]. As a result, research into faster (less computationally expensive) assembly and segmentation algorithms is continuing to take place. Additionally, researchers are investigating the use of graphical processing units (GPUs) for the capturing, segmentation, processing, analysis, and presentation of medical images [51, 96, 97].

7.1 Compression

Challenges with storage can be addressed through compression schemes. While image/video compression algorithms developed over the past 50 years are well established, they are primarily driven by human visual perception. For this reason, among others, standards such as the Digital Imaging and Communications in Medicine (DICOM) have adopted image formats with lossless compression capabilities [98]; currently, the JPEG2000 format has been chosen [99] for these requirements.

Whether the above account will be compatible for bioprinting is still an unanswered question and optimization schemes are continuously being developed. However, researchers must keep in mind that compression via lossy formats can result in artifacts. Considering that bioprinting requires clean information at the cellular and tissue scales, employed lossy compression schemes must take this into account. Even the use of the well-established DCT transform for image compression [100] can result in localized smoothing, which may result in obstructions when printing critical tissue areas and therefore organs. Overall, it is important for bioprinting researchers to be aware of such issues and gain an understanding of all the factors involved before printing begins.

7.2 Slicing

Discussion within the previous sections reviewed the processes in (a) taking a 3D object (human), (b) representing it in terms of successive 2D scans, (c) extracting a sub-object of interest (organ), and (d) reconstructing it virtually to achieve a 3D model. Slicing is the next stage of the bioprinting process (Fig. 1) whereby the 3D model is separated back into distinct 2D sections to be transferred to a bioplotter for realization in a layer by layer fashion. Typically, the information is represented as a series of commands interpretable by the Numerical Control (NC) system (bioprinter). While there are many forms of this, the widely adopted standard is G-Code [101]. This code essentially instructs the printer what to do, addressing timing and form in the process.

It is important to note that the slicing algorithm must contain information related to a printer's capabilities in addition to the material(s) prior to printing. Information such as (a) material (s) properties, (b) extrusion/deposition method, (c) number of extruders and mechanism thereof, and (d) spatial awareness, in addition to a number of other characteristics of the printer itself. Once known, in addition to established knowledge of how bioinks behave, the slicer can build a virtual representation of how the replicated tissue/organ will look during manufacturing. Considering that commercial printers use thermoplastics, such materials have been studied extensively making it easy to develop a manufacturing model. However, thermoplastics are relatively simple when compared with the dynamic space of bioinks.

In place of delving into the fundamentals of slicing technology, other published documents have broached this subject before [102]. For slicing bio systems, it must be noted that the current slicing technologies are not equipped, as yet, to address the complexities and characteristics of biological materials. Current software can supply supporting structures and deposition of cell-laden materials is achievable, as evident from the literature. However, additional research must take place in order to handle the full complexities associated with biomaterials.

8 Perspective and Challenges

It is clear from the literature that 3D bioprinting technology is still in its infancy in terms of the technological and biological requirements. However, since its inception great progress has been made toward printing functional tissues and therefore worthy of further investigation. Throughout this voyage, a number of challenges have presented themselves to achieve the overarching goal of fully functional organs.

A "Grand Challenge" for bioprinting researchers therefore is to develop the capabilities for fabricating fully operational and long life cell structures more extensively than is now taking place, with an ultimate goal of printing entire organs. Reaching such a goal will require an epistemological multidisciplinary approach involving expertise from engineers, biologists, clinicians, radiographers, ethicists, computer scientists, physicians, and mathematicians [103]. Engineers, for example, face the challenges ahead: fast paced printing process, nozzle and cartridge design, printing resolutions, reduce/eliminate clogging problems, and creating the conditions under which cell viability can be maximized. According to Malda, the lack of suitable bioinks is hampering the progress of the field, and poses a challenge for biochemists and material scientists [7]. Progress in the technological domain is evident through commercial printers, while more work is required on the biological side.

While a multidisciplinary approach is almost mandatory for the future success of this field, teething issues can result stemming from misunderstandings between the fields. For example, it has been found that when designing a product, disagreements between the design and manufacturing stages can be very costly [104–106]. Often complexities in the design can result in challenges for the manufacturer and in many cases, it may not be achievable at all or with the available resources. Learning from the manufacturing industry in this case will allow for more integrated solutions [107]. Given that organs, and scans thereof, are inherently complex in nature, it is expected that manufacturing limitations will be encountered during the printing stage. Such challenges are inevitable and will require experts working together in order to solve these issues when the understanding and technology are at a stage to do so.

The road ahead is paved with numerous challenges for multiple disciplines and correspondingly the multidisciplinary domain. Integrating these approaches for the grand goal of organ printing is no easy task, and will require experts to learn from one another to break down the disciplinary devices. Overall, it is accepted that great challenges lie ahead, which in turn presents amazing opportunities within this field. However, without collaborative efforts the progress/success is unlikely or will require an extensive amount of time, beyond current estimates.

References

1. Bonassar LJ, Vacanti CA (1998) Tissue engineering: the first decade and beyond. J Cell Biochem 72(Suppl 30–31):297–303
2. Schulz A, Xu J, Zhu B et al (2017) Interactive design space exploration and optimization for CAD models. ACM Trans Graph 36:1–14
3. Kocovic P (2017) 3D printing and its impact on the production of fully functional components: emerging research and opportunities: emerging research and opportunities. IGI Global, Hershey
4. Larson MG, Bengzon F (2013) The finite element method: theory, implementation, and applications. Springer Science & Business Media, New York
5. Sarcar MMM, Mallikarjuna Rao K, Lalit Narayan K (2008) Computer aided design and manufacturing. PHI Learning Pvt. Ltd., New Delhi
6. Derby B (2012) Printing and prototyping of tissues and scaffolds. Science 338:921–926
7. Malda J, Visser J, Melchels FP et al (2013) 25th anniversary article: engineering hydrogels for biofabrication. Adv Mater 25:5011–5028
8. Ozbolat IT, Hospodiuk M (2016) Current advances and future perspectives in extrusion-based bioprinting. Biomaterials 76:321–343
9. Murphy SV, Atala A (2014) 3D bioprinting of tissues and organs. Nat Biotechnol 32:773–785
10. Sheldon DF, McTaggart W (1986) CAD/-CAM: computer-aided design and manufacturing. Comput-Aided Eng J 3:34
11. Elanchezhian C, Shanmuga Sundar G (2007) Computer aided manufacturing. Firewall Media, New Delhi
12. Sutherland IE (1963) Sketchpad. In: Proceedings of the May 21–23, 1963, spring joint computer conference on—AFIPS '63 (Spring)
13. Sutherland IE (1964) Sketchpad a man-machine graphical communication system. Simulation 2:R–3–R–20
14. Williamson M (1986) The impact of CAD on aerospace design. Aircr Eng Aerosp Technol 58:17–19
15. Ross A, Loomis HH (1978) Computer Aided Design of Microprocessor-Based Systems. In: 15th Design Automation Conference
16. Trivedi AV (1988) Impact of robotics and CAD/CAM on an industrial technology curriculum. In: Robotics and factories of the future '87. pp 803–806
17. Krouse JK (1982) What every engineer should know about computer-aided design and computer-aided manufacturing: the CAD/CAM revolution. CRC Press, Boca Raton
18. Society of Manufacturing Engineers (1975) CAD/CAM and the computer revolution: selected papers from CAD/CAM I and CAD/CAM II
19. Kapos T, Ashy LM, Gallucci GO et al (2009) Computer-aided design and computer-assisted manufacturing in prosthetic implant dentistry. Int J Oral Maxillofac Implants 24 (Suppl):110–117
20. Parkash H (2016) Digital dentistry: unraveling the mysteries of computer-aided design computer-aided manufacturing in prosthodontic rehabilitation. Contemp Clin Dent 7:289
21. Sajjad A (2016) Computer-assisted design/computer-assisted manufacturing systems: a

revolution in restorative dentistry. J Indian Prosthodont Soc 16:96–99

22. Finne H (1988) CAD/CAM and social science in Scandinavia. In: Social science research on CAD/CAM. pp 10–26

23. Adler SW (2000) The Revolution of CAD/-CAM in the Casting of Fine Jewelry

24. Hull CW (1984) Apparatus for production of three-dimensional objects by stereolithography. US Patent

25. Bandyopadhyay A, Bose S (2015) Additive manufacturing. CRC Press, Boca Raton

26. Khorram Niaki M, Niaki MK, Nonino F (2017) What is additive manufacturing? Additive systems, processes and materials. In: springer series in Adv Manuf. pp 1–35

27. Wimpenny DI, Pandey PM, Jyothish Kumar L (2016) Advances in 3D printing & additive manufacturing technologies. Springer, New York

28. Mannoor MS, Jiang Z, James T et al (2013) 3D printed bionic ears. Nano Lett 13:2634–2639

29. Koprnicky J, Najman P, Safka J (2017) 3D printed bionic prosthetic hands. In: 2017 IEEE International Workshop of Electronics, Control, Measurement, Signals and their Application to Mechatronics (ECMSM)

30. Saunders CG, Foort J, Bannon M et al (1985) Computer aided design of prosthetic sockets for below-knee amputees. Prosthetics Orthot Int 9:17–22

31. Baynes S (2016) Printable Prosthetics: The Design of a 3D Printed Swimming Prosthesis: a Thesis Submitted to the Victoria University of Wellington in Fulfilment of the Requirements for the Degree of Master of Design

32. Langer R, Vacanti J (1993) Tissue engineering. Science 260:920–926

33. Yang S, Leong KF, Du Z, Chua CK (2001) The design of scaffolds for use in tissue engineering. Part I. traditional factors. Tissue Eng 7:679–689

34. Loh QL, Choong C (2013) Three-dimensional scaffolds for tissue engineering applications: role of porosity and pore size. Tissue Eng Part B Rev 19:485–502

35. An J, Teoh JEM, Suntornnond R, Chua CK (2015) Design and 3D printing of scaffolds and tissues. Proc Est Acad Sci Eng 1:261–268

36. Bártolo PJ, Chua CK, Almeida HA et al (2009) Biomanufacturing for tissue engineering: present and future trends. Virtual Phys Prototyp 4:203–216

37. Leong K, Chua C, Sudarmadji N, Yeong W (2008) Engineering functionally graded tissue engineering scaffolds. J Mech Behav Biomed Mater 1:140–152

38. Mould RF (2018) A century of X-rays and radioactivity in medicine: with emphasis on photographic records of the early years. Routledge, Philadelphia

39. Suetens P (2017) Fundamentals of medical imaging. Cambridge University Press, Cambridge

40. Bertoldi S, Farè S, Tanzi MC (2011) Assessment of scaffold porosity: the new route of micro-CT. J Appl Biomater Biomech 9:165–175

41. Herman GT (2009) Fundamentals of computerized tomography: image reconstruction from projections. Springer, New York

42. Mankovich NJ, Samson D, Pratt W et al (1994) Surgical planning using three-dimensional imaging and computer modeling. Otolaryngol Clin N Am 27:875–889

43. Hall EJ, Brenner DJ (2008) Cancer risks from diagnostic radiology. Br J Radiol 81:362–378

44. Vlaardingerbroek MT, Boer JA (2013) Magnetic resonance imaging: theory and practice. Springer Science & Business Media, New York

45. Kim S-G, Bandettini PA (2006) Principles of functional MRI. In: Functional MRI. pp 3–23

46. Thurston RN, Papadakis EP, Pierce AD (1998) Ultrasonic instruments and devices II: reference for modern instrumentation, techniques, and technology. Elsevier, Amsterdam

47. Bierig SM, Michelle Bierig S, Jones A (2009) Accuracy and cost comparison of ultrasound versus alternative imaging modalities, including CT, MR, PET, and angiography. J Diagn Med Sonogr 25:138–144

48. Thomenius KE (2009) Miniaturization of ultrasound scanners. Ultrasound Clin 4:385–389

49. Hangiandreou NJ (2003) AAPM/RSNA physics tutorial for residents: topics in US. Radiographics 23:1019–1033

50. Zhou Y (2016) The application of ultrasound in 3D bio-printing. Molecules 21:pii: E590. https://doi.org/10.3390/molecules21050590

51. Eklund A, Dufort P, Forsberg D, LaConte SM (2013) Medical image processing on the GPU–past, present and future. Med Image Anal 17:1073–1094

52. Hunt BR (1973) The application of constrained least squares estimation to image restoration by digital computer. IEEE Trans Comput C-22:805–812

53. Perona P, Malik J (1990) Scale-space and edge detection using anisotropic diffusion. IEEE Trans Pattern Anal Mach Intell 12:629–639

54. Granlund GH, Knutsson H (1995) Signal processing for computer vision

55. Elad M (2002) On the origin of the bilateral filter and ways to improve it. IEEE Trans Image Process 11:1141–1151

56. Coupe P, Yger P, Prima S et al (2008) An optimized blockwise nonlocal means denoising filter for 3-D magnetic resonance images. IEEE Trans Med Imaging 27:425–441

57. IEEE (2013) PET/CT image denoising and segmentation based on a multi observation and a multi scale Markov tree model. In: 2013 IEEE Nuclear Science Symposium and Medical Imaging Conference (2013 NSS/MIC)

58. Xu Z, Bagci U, Seidel J et al (2014) Segmentation based denoising of PET images: an iterative approach via regional means and affinity propagation. Med Image Comput Comput Assist Interv 17:698–705

59. Liu RW, Shi L, Huang W et al (2014) Generalized total variation-based MRI Rician denoising model with spatially adaptive regularization parameters. Magn Reson Imaging 32:702–720

60. Yang J, Fan J, Ai D et al (2016) Local statistics and non-local mean filter for speckle noise reduction in medical ultrasound image. Neurocomputing 195:88–95

61. Shapiro LG, Stockman GC (2001) Computer vision

62. Zhang Y-J Image Segmentation in the Last 40 Years. In: Encyclopedia of Information Science and Technology, Second Edition. pp 1818–1823

63. Roberts LG (1963) Machine perception of three-dimensional solids

64. Brice CR, Fennema CL (1970) Scene analysis using regions. Artif Intell 1:205–226

65. Zaitoun NM, Aqel MJ (2015) Survey on image segmentation techniques. Procedia Comput Sci 65:797–806

66. Barbieri AL, de Arruda GF, Rodrigues FA et al (2011) An entropy-based approach to automatic image segmentation of satellite images. Phys A: Statis Mechan Appl 390:512–518

67. Mejia-Inigo R, Barilla-Perez ME, Montes-Venegas HA (2009) Color-based texture image segmentation for vehicle detection. In: 2009 6th International Conference on Electrical Engineering, Computing Science and Automatic Control (CCE)

68. Bosch M, Zhu F, Delp EJ (2011) Segmentation-based video compression using texture and motion models. IEEE J Sel Top Signal Process 5:1366–1377

69. Smith SM (2002) Fast robust automated brain extraction. Hum Brain Mapp 17:143–155

70. Yazdani S, Yusof R, Karimian A et al (2015) Image segmentation methods and applications in MRI brain images. IETE Tech Rev 32:413–427

71. Petitjean C, Dacher J-N (2011) A review of segmentation methods in short axis cardiac MR images. Med Image Anal 15:169–184

72. Slomka PJ, Dey D, Sitek A et al (2017) Cardiac imaging: working towards fully-automated machine analysis & interpretation. Expert Rev Med Devices 14:197–212

73. Xie J, Jiang Y, Tsui H-T (2005) Segmentation of kidney from ultrasound images based on texture and shape priors. IEEE Trans Med Imaging 24:45–57

74. Mharib AM, Ramli AR, Mashohor S, Mahmood RB (2011) Survey on liver CT image segmentation methods. Artif Intell Rev 37:83–95

75. Moccia S, De Momi E, El Hadji S, Mattos LS (2018) Blood vessel segmentation algorithms - review of methods, datasets and evaluation metrics. Comput Methods Prog Biomed 158:71–91

76. Dogdas B, Stout D, Chatziioannou AF, Leahy RM (2007) Digimouse: a 3D whole body mouse atlas from CT and cryosection data. Phys Med Biol 52:577–587

77. Lavdas I, Glocker B, Kamnitsas K et al (2017) Fully automatic, multiorgan segmentation in normal whole body magnetic resonance imaging (MRI), using classification forests (CFs), convolutional neural networks (CNNs), and a multi-atlas (MA) approach. Med Phys 44:5210–5220

78. Ackerman MJ (2016) The visible human project®: From body to bits. In: 2016 38th Annual International Conference of the IEEE Engineering in Medicine and Biology Society (EMBC)

79. Danchin A, Médigue C, Gascuel O et al (1991) From data banks to data bases. Res Microbiol 142:913–916

80. Petrila T, Trif D (2006) Basics of fluid mechanics and introduction to computational fluid dynamics. Springer Science & Business Media, New York

81. Sun W, Starly B, Nam J, Darling A (2005) Bio-CAD modeling and its applications in

computer-aided tissue engineering. Comput Aided Des Appl 37:1097–1114

82. Lacroix D, Chateau A, Ginebra M-P, Planell JA (2006) Micro-finite element models of bone tissue-engineering scaffolds. Biomaterials 27:5326–5334

83. Chai X, van Herk M, van de Kamer JB et al (2011) Finite element based bladder modeling for image-guided radiotherapy of bladder cancer. Med Phys 38:142–150

84. Noritomi P, Xavier T, Silva J (2011) A comparison between BioCAD and some known methods for finite element model generation. In: Innovative developments in virtual and physical prototyping. pp 685–690

85. Anagnostakis A, Pappas A, Sucaet Y, Waelput W (2014) Digital pathology data brokerage: a standard recommendation for complex digital pathology information web-services. Anal Cell Pathol 2014:1–2

86. Sucaet Y, Waelput W (2014) Digital pathology. Springer, New York

87. Paganelli C, Summers P, Gianoli C et al (2017) A tool for validating MRI-guided strategies: a digital breathing CT/MRI phantom of the abdominal site. Med Biol Eng Comput 55:2001–2014

88. Wang C, Yin F-F, Segars WP et al (2017) Development of a computerized 4-D MRI phantom for liver motion study. Technol Cancer Res Treat:1533034617723753

89. Veress AI, Segars WP, Weiss JA et al (2006) Normal and pathological NCAT image and phantom data based on physiologically realistic left ventricle finite-element models. IEEE Trans Med Imaging 25:1604–1616

90. Mukai N, Takahashi T, Chang Y (2016) Particle-based Simulation on Aortic Valve Behavior with CG Model Generated from CT. In: Proceedings of the 11th Joint Conference on Computer Vision, Imaging and Computer Graphics Theory and Applications

91. Bolwin K, Czekalla B, Frohwein LJ et al (2018) Anthropomorphic thorax phantom for cardio-respiratory motion simulation in tomographic imaging. Phys Med Biol 63:035009

92. Dao TT, Tho M-CHB (2014) Biomechanics of the musculoskeletal system: modeling of data uncertainty and knowledge. Wiley, Hoboken

93. Kajiyama K (2016) Domestic market trend for medical imaging and radiological system. Nihon Hoshasen Gijutsu Gakkai Zasshi 72:717–719

94. Scholl I, Aach T, Deserno TM, Kuhlen T (2010) Challenges of medical image processing. Comput Sci Res Dev 26:5–13

95. Suganya R, Rajaram S, Sheik Abdullah A (2018) Big data in medical image processing. CRC Press, Boca Raton

96. Després P, Jia X (2017) A review of GPU-based medical image reconstruction. Phys Med 42:76–92

97. Smistad E, Falch TL, Bozorgi M et al (2015) Medical image segmentation on GPUs—a comprehensive review. Med Image Anal 20:1–18

98. Pianykh OS (2009) Digital imaging and Communications in Medicine (DICOM): a practical introduction and survival guide. Springer Science & Business Media, New York

99. Taubman D, Marcellin M (2012) JPEG2000 image compression fundamentals, standards and practice: image compression fundamentals, standards and practice. Springer Science & Business Media, New York

100. Fryza T (2006) Improving quality of video signals encoded by 3D DCT transform. In: Proceedings ELMAR 2006

101. Systems Management Council Interchangeable Variable Block Data Format for Positioning, Contouring, and Contouring/Positioning Numerically Controlled Machines

102. Evans B (2012) Practical 3D printers: the science and art of 3D printing. Apress, New York

103. Munaz A, Vadivelu RK, St. John J et al (2016) Three-dimensional printing of biological matters. J Sci Adv Mat Dev 1:1–17

104. Chang K-H (2013) Product manufacturing and cost estimating using CAD/CAE: the computer aided engineering design series. Academic Press, Cambridge

105. Locascio A (2001) Manufacturing cost modeling for product design. In: Information-based manufacturing. pp 315–325

106. Ulrich KT (2015) Does product design really determine 80% of manufacturing cost? (Classic Reprint)

107. CAD'15 (2015) Knowledge integration in CAD-CAM process chain. In: CAD'15

Chapter 4

Ethics and Policy for Bioprinting

Eliza Goddard and Susan Dodds

Abstract

3D bioprinting involves engineering live cells into a 3D structure, using a 3D printer to print cells, often together with a compatible 3D scaffold. 3D-printed cells and tissues may be used for a range of purposes including medical research, in vitro drug testing, and in vivo transplantation. The inclusion of living cells and biomaterials in the 3D printing process raises ethical, policy, and regulatory issues at each stage of the bioprinting process that include the source of cells and materials, stability and biocompatibility of cells and materials, disposal of 3D-printed materials, intended use, and long-term effects. This chapter focuses on the ethical issues that arise from 3D bioprinting in the lab—from consideration of the source of cells and materials, ensuring their quality and safety, through to testing of bioprinted materials in animal and human trials. It also provides guidance on where to seek information concerning appropriate regulatory frameworks and guidelines, including on classification and patenting of 3D-bioprinted materials, and identifies regulatory gaps that deserve attention.

Key words Human research ethics, Animal research ethics, Governance, Regulation, Bioethics, 3D Bioprinting, Stem cells

---

1 Introduction

3D bioprinting involves the use of 3D printers to engineer live cells into a 3D structure for purposes that include increased understanding of biological processes or conditions and improving patient well-being and health. The addition of living cells within the printing process engenders technical challenges related to cell viability and proliferation—including finding biocompatible materials to serve as scaffolds for cells and that protect cells during the printing process while achieving mechanical and functional properties for tissue constructs [1: page 778]. The inclusion of living cells within the printing process also raises ethical issues related to the source of the living cells and the intended use of the bioprinted materials, policy issues related to protecting human health and safety, and has begun the demand for regulatory frameworks to address regulatory challenges, which researchers will need to anticipate and adequately respond to in developing 3D printing techniques and applications.

Jeremy M. Crook (ed.), *3D Bioprinting: Principles and Protocols*, Methods in Molecular Biology, vol. 2140,
https://doi.org/10.1007/978-1-0716-0520-2_4, © Springer Science+Business Media, LLC, part of Springer Nature 2020

The typical process for 3D bioprinting of tissue involves three stages: preprinting, printing, and postprinting. The preprinting stage includes imaging, design approach, and selection of scaffold material and cells. The printing stage involves selection of bioprinting method. The postprinting stage focuses on the application of the bioprinted material, including maturation, implantation, and in vitro testing [1, 2]. In addition to technical challenges, ethical issues arise at each stage of the bioprinting process. The kinds of ethical issues that arise will depend, in part, on the purpose of research being undertaken and the application to which it is put, be it fundamental research (e.g., the study of cell behavior to understand processes of organisms); development of new human and/or animal models for in vitro diagnostics (e.g., pharmaceutical testing); or in vivo therapeutic applications (e.g., the implantation of 3D-bioprinted materials). Researchers are moral agents with ethical responsibilities at each stage of research, from the bench to clinical research [3]. Attention needs to be paid to research design, including assessment of the purpose and use of the research. Some of these issues are common to other research relevant to clinical care (such as accurate and appropriate recording of methods and findings).

Because of the novelty and breadth of potential applications, 3D bioprinting raises broad ethical questions, akin to other emerging technologies, including questions about justice in access to treatment [4] and the legitimate purposes of 3D bioprinting research, for example, treatment of conditions versus extending or enhancing human function [5]. While these broader issues are important, this chapter focuses on ethical and regulatory issues that researchers who make use of 3D bioprinting in the lab will need to take into account (*see* **Note 1**). Given the ultimate purpose of most 3D bioprinting is to advance human health, this chapter starts with a general discussion of the use of humans in research and the importance of research design. The following four sections discuss ethical and policy issues relating to (1) consideration of the source of cells and materials (human, nonhuman, and other), including issues related to donation and consent of human cells, (2) ensuring the quality and safety of cells and materials used in 3D bioprinting, (3) testing the safety and efficacy of 3D-bioprinted materials in animals and humans, and (4) classification and patenting of 3D-bioprinting technologies. Each section concludes with a suggested set of questions researchers might ask themselves to assist in orienting themselves toward the ethical, policy, and regulatory issues that arise from their research. These sections refer to policy and regulatory frameworks including the USA, Australia, European Union, and UK. Reference to specific regulation in this chapter is not intended to be exhaustive; rather, it is to indicate the types of issues that may arise and where to look in the regulations to address these.

Researchers should be mindful that there is no single regulatory regime that governs the entire bioprinting process, and that there are different regulatory approaches in different jurisdictions [6: page 441]. There is a range of regulations and guidelines or law that may be relevant to the technique and type of research. In part, this arises from the number of techniques that are used in the 3D bioprinting process, as well as the integration of materials from varied biological and synthetic origins in novel ways. Rapid advances in materials science and techniques, and the multiple applications to which 3D-bioprinted material is put, challenge regulatory frameworks to keep up [6: page 441]. As such, researchers need to attend to multiple forms of regulation, which may relate to the materials and the purpose or application of their research, as well as be mindful of potential regulatory gaps and anticipate ethical issues where there is a lack of clear guidance. As this is an area of rapid regulatory change, researchers should consult regularly with local research governance bodies for updates relevant to their research.

Questions to Consider:

- What is the purpose of the 3D bioprinting research being undertaken? For example:
    - Establishing a technique.
    - Testing the viability of cells.
    - Evaluating scaffold materials.
    - Developing diagnostic tools.
    - Developing therapeutics for use in humans.
    - Developing therapeutics for use in animals.
    - Animal testing of therapeutics for human use.
    - Animal testing of therapeutics for veterinary use.
- What is the source of cells used in the research? For example:
    - Nonhuman embryonic stem cells.
    - Human embryonic stem cells (hESCs).
    - Nonhuman-induced pluripotent stem cells (iPSCs).
    - Human iPSCs.
    - Autologous stem cells.
- Will the 3D-bioprinted materials be implanted into a living organism? If so:
    - Will the materials be implanted into the organism from which they derived (autologous)?
    - Will the materials be implanted into a different organism of the same species?

– Will the materials be implanted into a different species?

– Will 3D-bioprinted materials derived from a nonhuman species be implanted in a human?

• Have the materials being used as a scaffold for 3D bioprinting been tested and approved for use in humans? For veterinary purposes?

• Have the scaffold materials been tested for biocompatibility?

• If the 3D-printed materials are intended for use in a living organism, is the scaffold intended to be biodegradable or enduring? Have they been tested for biodegradability and long-term stability to ensure they do not cause scarring or infection; and are their interactions with other implants or therapeutics known?

A significant proportion of the research being undertaken with 3D bioprinting has as an ultimate aim the development of better diagnostics and therapies for human health. It is envisaged that developments in 3D bioprinting will allow for greater personalization of treatments by using patients' own cells to develop highly refined diagnostic tools—a "lab on a chip"—and integrated therapeutics printed into a patient's stem cells as part of the next phase of regenerative medicine, stem cell therapies and implanted targeted delivery. While 3D bioprinting is undertaken that focuses on proof of concept using nonhuman animal models, it is worth keeping in mind that if the ultimate purpose of a research program is to benefit human health, then that research needs to attend throughout to the ethical and regulatory significance of use of humans in research. This includes the justification of using people as research subjects where they may not be beneficiaries of the research, the ethical and regulatory requirements regarding minimizing risk, informed consent, privacy, and subsequent use of data from research.

Similarly, 3D bioprinting research involving nonhuman animals needs to be informed by awareness of the ethical concerns about the use of animals in research, regulatory requirements regarding use of animals, minimization of pain and suffering, ensuring appropriate stimulation and care, and the need to justify use of animals for research purposes. In the case of research involving xenografts (transfer of live cells between different species), there will often be additional regulatory controls to take into consideration.

2   Using Humans in Research

Humans may be used as the source of stem cells in 3D bioprinting (e.g., embryonic or adult stems cells) and may also be involved as participants in research on the effectiveness and reliability of 3D-bioprinted diagnostics and on the safety, effectiveness and side effects of 3D-bioprinted therapies. This means that ethical and regulatory issues may arise around the source of cells and the design and effectiveness of diagnostics, and therapeutic devices or treatments involving 3D bioprinting. The use of live cells derived from humans in the 3D bioprinting process, including uses with the ultimate purpose of developing clinical treatments, raises ethical issues about the use of humans in research.

Human research is research conducted with or about people or their data or tissue [7: page 3]. Research involving humans is an important step in ensuring advances to human health and well-being have a strong empirical basis and that treatments are safe, reliable, and effective, but research involving humans can also introduce significant risk of harm to participants in the research. Researchers have an ethical responsibility to minimize risks to participants through good practice with attention to ethical values which protect and respect human participants to ensure that the health, well-being, and autonomy of individual participants are adequately protected in the conduct of the research [8: page 50].

Countries have developed ethical guidelines or protocols for conducting and evaluating research involving human participants. Many of these guidelines derive in part from the ethical principles set out in the World Medical Associations' Declaration of Helsinki [9] and the Council for International Organizations of Medical Sciences (CIOMS) International Ethical Guidelines for Health Related Research Involving Humans [10]. Common to these guidelines are ethical principles relating to participant autonomy and informed consent, privacy, minimization of harm, and justice in the distribution of benefits and burdens of research. In Australia, for example, these are expressed as guiding ethical principles which should inform research: research merit and integrity, justice, beneficence, and respect [7]. Independent review boards also exist for ethical assessment and approval of proposals for conducting research involving human participants—in the USA, Institutional Review Boards (IRBs); in Canada, Research Ethics Boards (REBs); and in Australia, Human Research Ethics Committees (HRECs).

Historically, human research ethics guidelines and research ethics review committees have focused on the impact of research on human participants and on the groups affected by a particular area of research (e.g., social groups or populations sharing a particular condition). Some countries have legislation or guidelines concerning the extraction and use of cell lines derived from humans, some of which would be found in research ethics guidelines, but in some countries there would be separate legislation specifically addressing the extraction and use of cells and cell lines which may include information about the control of private information, required information for donors and advice about the agreed use of data derived from donated human cells (*see* Sect. 3.1 and Sect. 4 below). In addition to research ethics guidelines are guidelines about research on the human embryo, clones, chimeras, genetic material, and tissues [11, 12]; however, there is no single approach to regulation and guidelines on research involving human tissues, and as yet few countries have developed guidelines that specifically address 3D bioprinting of human cells or the use of 3D-bioprinted materials in research.

Before research commences researchers should anticipate the ethical impacts of their research. Attention should be paid to research design, including assessment of the purpose and use of the research, so as to reduce risk, ethical concerns, and unnecessary impacts on public health.

Questions to Ask About Research Involving Humans:
- Does the research involve collecting and using human cells or tissues? If so, consider the range of issues associated with collecting, storing, and using human specimens and tissues.

- Does the research involve humans as participants in research? If so, consider whether such participation is restricted to assessing the effectiveness and reliability of 3D-bioprinted diagnostics or also includes implanting human 3D-bioprinted human cells into people for therapeutic purposes.

- Does the research involve xenografts or implants derived from other species into humans? If so, refer to any specific regulation on xenotransplantation and use of animals in research.

3   Sources of Cells and Materials

A key concern in the preprinting stage is the selection of cells to be used in the printing process. When selecting cells, lab researchers are predominantly concerned with issues around proliferation and cell culture techniques, printability, degradation, structural and mechanical properties, and material biomimicry, as well as reliable cell sourcing [1]. In addition to these technical and practical concerns, the inclusion of live cells introduces ethical issues related to the source of these cells, which trigger different regulatory instruments. These are categorized below according to the source of the most commonly used live cells in 3D bioprinting—human, nonhuman animal, and other (viral/bacterial, plant)—as well as of biomaterials used in inks and as scaffolds. In addition, the section discusses issues related to donation and consent when using live cells derived from humans.

3.1   Human Sources of Cells

3D bioprinting may raise ethical concerns resulting from the use of human stem cells. Commonly used human stem cells in bioprinting are embryonic (hESCs), induced pluripotent (iPSCs), and adult stem cells [13]. hESCs are derived from the early human embryo, their use is controversial, with debate centering on the moral status of human embryos, as well as the interpretation of humanity and human dignity [14, 15]. Concerns have also been raised about the consent or exploitation of women who donate embryos for hESC research. Many jurisdictions (e.g., Australia and Canada) restrict researchers to using human embryos donated by couples or women for whom the embryos are "supernumerary," that is, the embryos were created through assisted reproductive technologies (ART) and are no longer required for the couple to achieve a pregnancy. In other jurisdictions, for example, the USA and India [16], women may choose to donate, or sell, gametes regardless of whether they have been created for reproductive purposes. Human embryo research and the use of embryos to derive hESCs are highly regulated in many countries. In Australia, the Research Involving Human Embryos Act 2002 (Cth) [17] regulates use of excess human embryos created through ART (*see* **Note 2**). In the European Union, regulation varies by member state; in the UK, the use of hESCs is regulated by the Human Fertilisation and Embryology Act 2008 [18]. Some patients consider it unethical to use 3D-bioprinted materials derived from hESCs and opposition evoked by hESCs is one of the disadvantages of using these types of cells [14: page 368]. In addition, innovations which fail the "morality test" may be excluded from patents (*see* Sect. 6 of this chapter).

Induced pluripotent stem cells (iPSCs) are not as controversial, nor are they subject to the same level of regulation as those created through hESCs, and increasingly they, or alternatives to hESCs, are used in 3D bioprinting. The use of iPSCs does raise issues around safety and quality assurance, given the risk of cell lines derived from adult cells leading to cancers, as well as ethical concerns relating to donation and procurement. Where research uses printed cells with the aim of implanting these in a patient as a therapy, there are concerns about risks to human health and safety related to cell stability and biocompatibility. The development of 3D-bioprinted therapies requires that consideration is given to more long term effects. Health and safety questions relate to how printed cells will behave when implanted and whether they could migrate elsewhere in the body or whether cells may mutate into cancer cells. While use of autologous cell lines may avoid regulatory constraints associated with use of allogenic or xenogenic cells, there may still be significant risks to patients, including risks of teratoma formation [19]. The use of allogeneic cells from donors, including from dead people, introduces risks related to potential immunological problems and the risk of disease transmission [14].

In addition to consideration of ethical issues related to the source of human cells, researchers who use donated human tissues in research should ensure that the cell lines they use have been obtained through an appropriate process which protects the interests of the donor and that the necessary processes for securing informed consent, privacy and relinquishments of ownership rights of the donated cell or tissue has occurred [20: page 14, 21]. Where cell lines are derived from human embryos, there will be protocols and regulation surrounding the donation of embryos for use in research, frequently governed by legislation on ARTs. If using adult stem cells, there will be different requirements for securing consent to obtain the cells and their use in research. In cases where the cells are extracted from tissue removed as part of a medical procedure, there may be no formal requirement to obtain consent for their use in research. However, some regulations do stress the need to obtain informed consent from a cell donor. In the Guidelines set out by CIOMS, Guideline 9 provides guidance on Individuals Capable of Giving Informed Consent [10]. In the EU, specific regulation related to donation of human tissues can be found in the Tissues and Cells Directive, Article 13 [22]; in the UK, the Human Tissue Act 2004, Part 1 [23]; and in Australia, in the Australian National Statement on Ethical Conduct in Human Research 2007, Chapter 2.2 [7]. An appropriate informed consent process minimizes the risk of harm and potential violation of interests, such as around privacy and control of medical information [20]. Further, as cells lines may survive the donors or be potentially used for projects unplanned at the time of tissue/cell procurement, donors should be informed as fully as possible about the future uses of their material and approval for use of research using this material should extend past the lifetime of the donor and/or research project [19].

In the case of donation of stem cells for research there may be problems involved in meeting the ideals of informed consent given that it may not be possible at the time of donation to foresee all future possible uses of cells lines and more general or "blanket" consent may fail to meet the standards expected of consent in research [14]. When a donor is recruited for therapeutic research, appropriate information about the 3D-bioprinted product and implantation process should be provided as part of the consent process, as well as details of any conflicts of interest and potential outcomes and adverse effects [24: page 288]. The privacy of the donor should be protected through anonymization of samples in scientific research [14: page 368]. Questions remain about whether donors should be paid or whether donation should be free or unpaid (for altruistic purposes), so as to facilitate scientific research and improve public health. Some propose that altruistic donation should be the ideal behind legislation regulating the collection of cells/tissues for research in regenerative medicine [14, 25]. Questions also remain about ownership of donated cells/tissue and whether the body can be subject to property rights. Whilst acknowledging transfer of ownership might undergird legal processes for paid donation, and the patenting of cells and cells lines; granting ownership or property rights in tissues and organs is often considered to violate human dignity and could lead to exploitation of vunerable people [14: page 369, 25].

3.2 Animal Sources of Cells

3D bioprinting can be undertaken with cell lines derived from animals to establish techniques for printing processes, including viability of cells printed with different processes and the interaction of different printed scaffold materials with cells. They may also be integrated into 3D-bioprinted therapies. The regulatory environment for deriving embryonic and adult stem cell lines from nonhumans is much less restrictive than in the case of human-derived cell lines; nonetheless, there may be animal welfare and biohazard controls to take into consideration (*see* Sect. 4 and Sect. 5.1 below).

Where cells derived from nonhuman animals are to be used for potential human therapies, researchers should attend to ethical concerns about the acceptability of using animal cells for research, as well as laws or regulation aimed at reducing the risk of transmission of disease across species or that prohibit the potential creation of chimeric (human–animal) organisms [26, 27]. Some patients may object to receiving therapies that involve the use of cells derived from animals on religious grounds (e.g., reservations of some Muslims and Jews regarding the use of porcine cells or tissue) or because of the appropriation of animals for human research and consumption [14, 20, 24]. The use of bioinks derived from animal cells raises risks including the risk of transfer of an infectious disease from animals to humans (zoonosis) and the potential immunological problems xenogeneic cells may cause [14, 20, 24] (*see* Sect. 4). Laws prohibiting the creation of chimeras may also apply to the introduction of human stem cells into nonhuman animals [28], although a number of countries permit the introduction of human cells into an animal embryo provided that the embryo is not then transferred into a human or animal for gestation [29].

3.3 Other Sources of Cells

If using cells derived from viruses or bacteria, health and safety are the primary ethical concerns, and researchers should be aware of practices around risk control, including those around occupational health and safety and the development of biosafety plans and attend to national and international biosafety hazard regulations. The World Health Organization Laboratory Biosafety Manual contains information about containment of biohazards and prevention and control of infection [30].

3.4 Bioink Components

Bioinks, in addition to live cells, involve the use of biomaterials, typically including additives (growth factors, chemicals, microfactors, others) and a supportive scaffold (hydrogels, synthetic or natural polymers). The inclusion of these materials, like that of live cells, raises ethical issues related to sourcing, environmental impact and biocompatibility. Scaffolds are not used in all 3D bioprinting approaches, however [31: page 189]. Hydrogels are polymers and can be made from a variety of components, including of natural and synthetic components. Whilst the advantages of using synthetic polymers include that they can be tailored, unlike natural polymers, their interactions and effects on cells have not been studied systematically [32]. Other concerns include poor biocompatibility, toxic degradation and loss of mechanical properties [1]. Components of natural polymers existing in extracellular matrix (ECM)—gelatin, collagen, laminin, and fibronectin, and other natural polymers such as alginate, chitosan and silk fibrin and hyaluronic acid, often isolated from animal or human tissues, are advantageous given their similarity to human ECM and inherent bioactivity [32]. Biomaterials derived from nonhuman organisms, such as gelatin (from porcine skin) or alginate (from seaweed) may carry risks, such as immunological responses and the risk of introducing pathogens [14]. For these reasons, researchers need to consider carefully the materials used in 3D bioprinting, paying particular attention to risks to humans or the environment from pathogens, and risks to any human participants in the research. In deciding which bioinks and scaffolds to use, there should be consideration given to the disposal of the materials in the printing process and in their eventual degradation and impacts on the environment.

Questions to Ask About the Source of Cells and Biomaterials:

- Does the research involve using cells and materials from human beings? If so, consider the range of issues associated with the source of stem cells—are they embryonic, induced pluripotent, or adult stem cells? In addition, consider the range of issues associated with the donation and collection of human cells, including informed consent, privacy, and relinquishments of ownership. Have you familiarized yourself with the regulations pertaining to donation and consent in your jurisdiction?

- Does the research involve using cells and materials from nonhuman animals? If so, consider the range of applicable law, regulation, and ethical issues, including whether there are potential risks related to zoonosis.

- Does the research involve using cells and materials from sources other than human and nonhuman animals? If so, consider health, safety, and environmental impacts.

- What is the source of the materials in the bioink and scaffold—synthetic, natural, or both? Do you intend to use the material in therapies? If so, consider issues related to biocompatibility, stability, and degradation.

4   Quality and Safety of Cells and Materials

In addition to consideration of the source of cells and materials selected for use in the bioprinting process, including consent for use of human cells, consideration must be paid to the collection, storage, and use of cells and materials to ensure their quality and safety. In order to ensure the quality and safety of cell and tissue material and prevent the introduction, transmission, and spread of communicable diseases, the early stage of donation, procurement and testing of cells is governed by regulations. These establish donor eligibility, good tissue practice and other procedures, including requirements for clean processing environments. Regulations concerning the legitimate origin and provenance of cell lines for use in research will vary by country and some regulations prevent certain cell lines being used in research. In addition to these measures, to ensure the quality and safety of cells and materials, attention should be paid, throughout the printing process, to good laboratory and manufacturing processes.

Most countries' biomedical research councils have adopted the Council of International Organizations of Medical Sciences (CIOMS) guidelines [10]. Guideline 11 governs the collection, storage, and use of biological materials and related data [10]. In the European Union, the EU Tissues and Cells Directive, sets standards of quality and safety for the donation, procurement, testing, processing, preservation, storage, and distribution of human tissues and cells [22]. In the UK, the Human Tissue Act 2004 regulates the removal, storage, and use and disposal of human tissue, overseen by the Human Tissue Authority (HTA) [23]. In the USA, therapies derived from human cells and tissues intended for transplantation in a human are regulated as human cells, tissues and cellular and tissue-based products (HCT/Ps). The Code of Federal Regulations (CFR) Parts 1270 and 1271 require tissue establishments to screen and test donors, to prepare and follow written procedures for the prevention of the spread of communicable disease, and to maintain records [33] (*see* **Note 3**). In Australia the use of human and animal cell lines in health and medical research is covered by guidelines and statements issued by the National Health and Medical Research Council [7]. Researchers should abide by the provisions in the National Statement on Ethical Conduct in Human Research 2007, Chapter 3.2. Human biospecimens in laboratory based research.

Sterilization of the environment, materials and machinery is important for ensuring cell safety and quality. The use of an aseptic environment extends to cell culture and biofabrication machinery [8]. 3D bioprinting must be performed within a regulatory framework of good laboratory practice (GLP) for standardized, optimized, and controlled tissue fabrication. GLPs vary by country; see the World Health Organization [34], the USA [35], and the European Union [36]. If you are printing material for use in humans, then adherence to the more rigorous good manufacturing practice (GMP) for clinical-product development is required (cf. Preface, Crook). When using 3D bioprinting techniques for biological applications, sterilization of the environment, materials and machinery is of absolute importance to prevent contamination of 3D-bioprinted products. GMP regulations vary by country and are overseen by regulatory agencies—see World Health Organization [37]; the USA [38], and European Union [39].

Questions to Ask About Ensuring the Quality and Safety of Cells:

- Are you aware of the range of issues related to the safety and quality of cells, including regulations around procurement, storage, disposal, and record-keeping?

- Are you familiar with good laboratory practice and conduct around ensuring clean processing environments?

- Are the tissue products intended for use in humans? Then, have you consulted the specific regulations? Are you familiar with regulations related to good manufacturing practice?

5   Testing the Safety and Efficacy of 3D-Bioprinted Material

Once the biomaterials have been printed, ethical and regulatory issues will arise with respect to testing the safety and efficacy of these materials. This may involve the use of animal and/or human trials. It is possible that future development of 3D bioprinting techniques, for example, using autologous human cells to print organ tissues, will avoid ethical concerns associated with organ transplant rejection and potentially reduce organ waiting lists and the illegal trade in human organs, or provide alternatives to drug and vaccine testing on animals and humans. Similarly, ex vivo testing on 3D-bioprinted animal or human tissues may reduce the dependence on use of humans and animals in in vivo tests. In the meantime, however, trials using animals and humans will remain the standard models for testing the safety and efficacy of novel materials and devices.

5.1   Animal Testing

Having secured a cell line and printed live cells into a 3D structure, researchers may seek to conduct tests to establish the reliability of diagnostics, the biocompatibility of the materials or the efficacy of a therapeutic application. Any animal study will be governed by regulations as discussed below. Where researchers are developing 3D-bioprinted materials for implants, researchers need to consider whether all the materials are safe to implant in the human body. They will need to assess the stability, biodegradability and biocompatibility of the materials, and test the behavior of cells and scaffold materials, including assessing cell migration and mutation, the potential for immunological reactions, the long-term health effects, including the potential for teratoma and cancer, the dislodgement and migration of implants, as well as cell and material stability or degradation. These questions are usually determined using a pathway starting with in vitro trials to establish whether the new

3D-bioprinted materials (and devices) are safe for use. These will then be followed by clinical trials on animals to establish both biocompatibility and effectiveness before being tested on humans.

The use of animals in research requires ethical justification (see, e.g., the Australian Code for the Care and Use of Animals for Research [40]). Since the latter half of the twentieth century, there has been increasing concern about the use of animals in medical experiments and research [41]. The recognition that animals experience discomfort and pain demands that the use of animals should be minimized and any unnecessary suffering should be avoided. Regulatory frameworks for the use of animals adopt the principles of replacing, reducing, and refining the use of animals in scientific research [40].

In most countries, laws, regulation, and guidelines addressing the care and use of laboratory animals require researchers to obtain approval for their research from animal research ethics and welfare committees to assess whether the use of animals in the proposed research is justified and appropriate. The UK has comprehensive regulation, writing its ethical framework into law by implementing The Animals (Scientific Procedures) Act 1986 [42], which requires experiments to be regulated by three licenses—a project license for the scientist in charge of the project, a certificate for the institution, and a personal license for each scientist or technician. In addition, all licensed establishments must have an Animal Welfare and Ethical Review Body (AWERB). In Australia, Animal Ethics Committees (AECs) oversee and have authority to approve the use of animals in research within research institutions. In the USA, animal testing on vertebrates is primarily regulated by the Animal Welfare Act of 1966 (AWA) [43] and the Animal Welfare Regulations [44]. In addition, the AWA requires institutions to maintain an Institutional Animal Care and Use Committee (IACUC) to ensure local compliance with the Act. In the EU, experiments on vertebrate animals (since 2013) are subject to Directive 210/63/EU on the protection of animals used for scientific purposes [45]. In Japan, the main law is the "Act on Humane Treatment and Management of Animals," which mandates a self-regulation system for animal experimentation. In addition, there are numerous voluntary guidelines provided by various organizations. Animal research in China is currently regulated and administratively managed according to national and provincial laws, regulations, guidelines, and standards [46: page 303].

Questions to Ask Concerning Animal Testing:

- Does your research involve testing on animals? If so, have you considered the range of issues related to reduction, replacement and refinement of the use of animals in research?

- Do you need to obtain ethics approval from an Animal Research Ethics Committee?
- Have you obtained the necessary licenses?

5.2   Human Trials

Once preliminary biocompatibility and safety tests on animals indicate that the 3D-bioprinted materials are not harmful and have been shown to be effective in principle, the safety and effectiveness of new diagnostics and therapeutics is then tested on humans in clinical trials. Approval to undertake clinical trials requires independent ethical review (in the USA, Institutional Review Boards [IRBs]; in Canada, Research Ethics Boards [REBs]; and in Australia, Human research Ethics Committees [HRECs]). Approval of a drug, device, or diagnostic tool for sale or use in clinical settings requires the additional registration and approval through national drug and therapeutic good regulators such as the US Food and Drug Administration (FDA) or the Australian Therapeutic Goods Administration (TGA) and through regulations such as the EU Medical Device Regulations [47] (*see* **Note 4**). In the development of 3D-bioprinted diagnostic devices, for example, a clinical trial of the device would seek to establish that it met an appropriate threshold of reliability initially, with later trials establishing, for example, that it was as well or better accepted by patients (in terms of discomfort or inconvenience), that it was no more costly and allowed for earlier or more fine-grained diagnosis of the relevant condition. The research ethics review process would be governed by principles designed to protect the interests of participants in research, while the overarching drugs and therapeutic goods regulation is designed to protect the wider group of people who may rely on the diagnostic device in clinical contexts. Reference should be made to human research regulations cited in Sect. 2. Some jurisdictions have separate regulations which pertain specifically to clinical trials. These should be considered alongside human research ethics guidelines.

Clinical trials are subject to various regulatory controls to ensure the safety of participants; however, the personalized nature of 3D-bioprinted products may challenge the use of randomized clinical trials as the primary means for testing new medical products and devices. At present in Australia, for example, there is no specific regulatory framework to guide the testing of 3D-bioprinted treatments in human patients. Traditionally, the safety and efficacy of an experimental treatment has been established through progressively larger samples of patients receiving either the experimental treatment or a placebo to establish a statistically significant effect of the experimental treatment. More frequently now the experimental treatment is compared with the most effective treatment currently available. However, where the experimental treatment involves a number of new variables such as a new surgical technique to implant 3D-printed devices as well as the possibility of using the patient's own cells as the source of the printed cells, it may be difficult to adapt the clinical trial methodology to provide meaningful evidence of safety and efficacy unless the complex array of novel interventions is broken down into more specific trials. To the degree that the novel treatment is individual to the patient, for example, using her cells and designed around the particularities of her organs or physical structure and involves assessing a very complex array of novel factors, including surgical methods, and the interaction of new materials with her biochemistry, as well as, conceivably, new drugs that are delivered in new ways, then the trials of the various elements may not be able to be assessed discretely [8: page 50]. New methods for evaluating 3D-bioprinted therapeutics for clinical approval will be needed which assess the complex array of techniques, materials, and functions, and which acknowledge the limits of generalizations based on clinical outcomes given the differences between individual patients.

Questions to Ask Concerning Human Testing:

- Does your research involve testing on humans? If so, do you need to obtain ethics approval from an Independent Review Board/Human Research Ethics Committee?

- How have you designed your trial to ensure the quality and significance of the findings while also protecting the interests of research participants?

6   Classification and Patenting of 3D-Bioprinted Products

6.1   Classification of 3D-Bioprinted Products and Devices

Where the focus of research is on the final translation of 3D-bioprinting research into clinical practice, a key issue in postprinting will be the classification of bioprinted products in order to assess the appropriate regulatory mechanism for approval of the product. Regulatory regimes have tended to define therapeutic products as single entities with respect to their use and functional aspects—for example, as a pharmaceutical, biologic, or medical device [48: page 2]. 3D-Bioprinted products and approaches, by combining components or sharing features from one or more of these categories, challenge the demarcations between regulated product categories. For example, a 3D-bioprinted product may combine cell laden tissue products with a device, potentially also with a drug delivery mechanism. Under some regulations, 3D-bioprinted products are often described as combination products—broadly defined as containing one or more regulated components.

In Australia, the Therapeutic Goods Administration (TGA) is responsible for regulating therapeutic goods under the therapeutic goods legislation. This legislation refers to classes of Medical

Devices and Biologicals, the former classified in the Therapeutic Goods (Medical Devices) Regulations 2002 [49] and the latter, Therapeutic Goods Regulations 1990 [50]. The TGA treats therapeutic products developed from 3D bioprinting as a "borderline" or "combination" treatment which is part medical device, part biological [51]. The Australian Regulatory Guidelines for Medical Devices [51] and Australian Regulatory Guidelines for Biologicals are also relevant [52]. In the EU, therapeutic products are classified as either medicinal products or medical devices; both the Advanced Therapy Medicinal Products (ATMP) regulation [53] and Medical Device Regulations [47] are relevant. The ATMP Regulation classifies tissue-based or cell-based products as medicinal products, including gene therapies, cell therapies, and tissue engineering [6, 54]. In the USA, the FDA classifies therapeutic products as pharmaceuticals, biologics and medical devices, each of which is overseen by a different section of the FDA—the Center for Drug Evaluation and Research; Center for Biologics Evaluation and Research; and, Center for Devices and Radiological Health, respectively. 3D-bioprinted therapeutic and diagnostic products may be considered by the FDA as a "combination product," when it includes medical devices that combine biological materials, medical devices, and/or drugs of different compositions [55]. Given the rapid progress and novel applications of 3D-bioprinting technologies, many relevant regulations are either currently under review or will be revised soon. Researchers should be aware of the rapidly evolving and changeable regulatory framework for classification of 3D-bioprinting technologies.

Software and hardware set ups for 3D bioprinting may be regarded as medical devices (that would then fall within the relevant regulations, for example, the EU Medical Device Regulations). If so, software and hardware setups will need certification before being released on the market [6]. This would mean software would fall subject to safety and performance requirements and also raises data protection issues.

In addition to concerns about classification, combination products raise issues related to consumer safety and protection. A combination product will command a high level of regulatory scrutiny, particularly for the manufacturer that makes multiple constituent parts. The key regulatory focus areas for combination products are: the process from design to manufacture; software system chain control and validation; and potential variation in critical quality attributes of the final manufactured product [48]. The mass digitization of stereolithography (STL) files pose risks to the regulatory framework of consumer safety, product liability (quality control), data protection, confidentiality, and safety [6, 48, 54].

6.2 Intellectual Property and Patenting

Researchers should be aware of the relevant Intellectual Property (IP) framework and the requirements to protect their interests prior to publishing their research or moving to clinical trials. Relevant areas of focus include: machines, methods, materials, processes, and products (e.g., kidney cell). There is uncertainty within the IP legal landscape for bioprinting including uncertainty over what aspects can be protected, as well as which aspects of patentability and copyright should not be protected for ethical or public policy reasons.

The law around patenting in the 3D bioprinting area is evolving. The potential for patenting of 3D bioprinting technologies is a contentious matter and a topic of contemporary legal and ethical discussions [54, 56, 57]. Patentable subject matter is that which is new (novel), useful (industrial application or utility) and contains an inventive aspect or step. Patents could apply to methods of 3D bioprinting and to 3D-bioprinted products, including bioprinters, bioprinted materials, as well as the fabrication and postproduction maturation processes. Given the customized nature of 3D bioprinting, patents over methods or processes are likely to predominate. In contrast, given the personalized or bespoke nature of 3D treatments, patenting of 3D-bioprinted products may be less useful and so less likely. However, whether innovative methods and products used in biomedical inventions constitute patentable subject matter is not clear. For example, many jurisdictions exclude patenting of human beings or tissue. In some jurisdictions, patents are not extended to products that are identical to the natural element, although materials isolated from the human body or otherwise produced by a technical process may constitute a patentable product. With reference to these directives 3D-bioprinted tissues using hydrogel and a scaffold may be patentable but a 3D-printed organ produced by 3D scanning and computer-aided design (CAD) files may not [54: page 288]. Patents have been granted for 3D bioprinting technologies (including for processes and products, for example, "Multilayered Vascular Tubes" by Organovo) [54: page 286].

However, given the natural elements of the product itself, many argue that IP law should not apply to 3D-bioprinted products; they argue these technologies should not be protected. Whilst patents may protect the inventions of researchers (and investment by companies), patents applied to 3D-bioprinting technologies would prevent others from making, using or selling patented methods and products. This raises ethical issues related to access and benefit sharing of these technologies, they may restrict patient access to medical treatment and potentially hinder innovation [58]. As in the area of genetic technologies and diagnostic testing, patent granting has had monopolistic effects which have widened gaps in health [54: page 283].

In the interests of access and equity to health technologies some countries have introduced exclusions from patentability, including based on a "morality test" and whether methods constitute medical treatment or therapy. The European Patent Commission, for example, excludes the commercial exploitation of an invention, if such activities are deemed to offend against the standards of morality [54: page 287, 59]. Existing Patents laws in India exclude patents which are contrary to public order or morality [60]. Exemptions for methods of medical treatment or therapy may include diagnostic methods, or may not, depending on how these are defined [54: page 293] and whether the diagnostic method is in vitro or in vivo (currently under UK patent law, the former is patentable). In the USA, patents for medical treatment have been more commonplace, although therapeutic exemptions exist and, in recent years, the approach in determining the scope of patentability has been more strictly adopted [54: pages 293–294].

To develop an equitable approach to innovation and benefit sharing of access to health technologies, alternative governance models to patents have been canvassed [54, 58]. These include the adoption of a portfolio approach to innovation [54]. Others advocate limited patents over certain elements of bioprinting.

In contrast to 3D-bioprinted products or processes, copyright protection might be more applicable to developed software rather than patent protection. Copyrights would protect the CAD-CAM (Computer Aided Manufacturing) files used in 3D bioprinting for scanning, manufacturing and bioprinter control. Although, copyright laws may be inadequate; trademark law for hardware and software protection and trade secret protection may present some avenues for bioprinting rights. Similar ethical issues to those of patenting—concerns related to equity and access—are raised with respect to copyrighting 3D bioprinting software. For example, it is argued that if a software is developed that allows for printing functional organs, for the benefit of society, its availability should not be restricted.

Given the evolving legal landscape, as well as the specificities, researchers should liaise with the IP and patenting office at their institution or research facility.

Questions to Ask Concerning Bioprinted Products (and Processes):
- Does your research involve the development of a bioprinted product that will require regulatory approval before it can be used clinically?
- Have you considered the regulatory requirements early in the design of your 3D-bioprinted devices or therapy?
- Will your research innovate a 3D bioprinting product or process? Have you considered the IP environment and requirements for asserting IP or for commercializing your research?

7   Conclusion

3D bioprinting presents real possibilities for understanding, diagnosing, and developing new therapies. 3D-printed cells and living structures are being used in medical research, in vitro drug testing and as in vivo transplant materials. The incorporation of live cells in the printing process raises ethical, policy, and regulatory issues at all stages of the bioprinting process, relative to the purposes of research and application of the research. This chapter focuses on the ethical and regulatory issues that arise in the 3D bioprinting process seeking to assist researchers in the lab to consider the ethical and regulatory issues at each stage. This chapter identifies ethical issues in relation to the use of humans and animals in research; the source of materials and cells; the quality and safety of cell and materials; donation and collection of human cells; the testing of materials, and animal and clinical trials. It also identifies regulatory issues related to the classification of 3D-bioprinted materials and products, and intellectual property and patenting.

8   Notes

1. This chapter does not discuss approaches to bioprinting, nor particular printing techniques; these are discussed in Part B of this volume.

2. Research involving the use of human embryonic stem cell lines, including human embryonic stem cell lines that have been imported into Australia, is not subject to specific regulation in Australia.

3. Bioprinted tissues typically used in research do not require FDA approval during animal and in vitro testing because they are not intended for use in humans.

4. New Regulations on medical devices were adopted in 2017, replacing existing directives. The new rules will apply after a transitional period which ends on 26 May 2020.

Acknowledgments

The authors wish to acknowledge funding from the Australian Research Council (ARC) Centre of Excellence Scheme (CE140100012).

References

1. Murphy SV, Atala A (2014) 3D bioprinting of tissues and organs. Nat Biotechnol 32 (8):773–785
2. Shafiee A, Atala A (2016) Printing technologies for medical applications. Trends Mol Med 22(3):254–265. https://doi.org/10.1016/j.molmed.2016.01.003
3. Baker HB, McQulling JP, King NM (2016) Ethical considerations in tissue engineering research: case studies in translation. Methods 99:135–144. https://doi.org/10.1016/j.ymeth.2015.08.010
4. Tuckson RV, Newcomer L, De Sa JM (2013) Accessing genomic medicine: affordability, diffusion, and disparities. JAMA 309 (14):1469–1470
5. Allhoff F, Lin P, Steinberg J (2011) Ethics of human enhancement: an executive summary. Sci Eng Ethics 17:201–212. https://doi.org/10.1007/s11948-009-9191-9
6. Li P, Faulkner A (2017) 3D bioprinting regulations: a UK/EU perspective. Eur J Risk Regulat 8(2):441–447. https://doi.org/10.1017/err.2017.19
7. National Health and Medical Research Council, Australian Research Council and Universities Australia (2007, Updated 2018) National Statement on ethical conduct in human research. Commonwealth of Australia, Canberra. https://www.nhmrc.gov.au/about-us/publications/national-statement-ethical-conduct-human-research-2007-updated-2018
8. Wallace G, Cornock R, Connell C et al (2014) 3D bioprinting: printing parts for bodies. ARC Centre of Excellence for Electromaterials Science, Australia
9. World Medical Association (1964, updated 2013) World medical association declaration of Helsinki—ethical principles for medical research involving human subjects, J Am Med Assoc 310(20):2191–2194. https://doi.org/10.1001/jama.2013.281053. https://www.wma.net/policies-post/wma-declaration-of-helsinki-ethical-principles-for-medical-research-involving-human-subjects/
10. Council for International Organizations of Medical Sciences and World Health Organization (1982, Updated 2016) international ethical guidelines for Health related research involving humans. 4th edition. CIOMS, Geneva. https://cioms.ch/wp-content/uploads/2017/01/WEB-CIOMS-EthicalGuidelines.pdf. Accessed 9 Aug 2019
11. The Academy of Medical Sciences (2011) Animals Containing Human Material. https:// acmedsci.ac.uk/file-download/35228-Animalsc.pdf. Accessed 9 Aug 2019
12. Prohibition of Human Cloning for Reproduction Act 2002 (Cth). Australian Government. https://www.legislation.gov.au/Series/C2004A01081
13. Leberfinger AN, Ravnic DJ, Dhawan A et al (2017) Concise review: bioprinting of stem cells for translatable tissue fabrication. Stem Cells Transl Med 6:1940–1948
14. de Vries RBM, Oerlemans A, Trommelmans L et al (2008) Ethical aspects of tissue engineering: a review. Tissue Eng Part B Rev 14 (4):367–375
15. de Wert G, Mummery C (2003) Human embryonic stem cells: research, ethics and policy. Hum Reprod 18(4):672–682
16. Klitzman R, Sauer MV (2015) Creating and selling embryos for "donation": ethical challenges. Am J Obstet Gynecol 212 (2):167–170. https://doi.org/10.1016/j.ajog.2014.10.1094
17. Research Involving Human Embryos Act 2002 (Cth). Australian Government. https://www.legislation.gov.au/Series/C2004A01082
18. Advanced Human Fertilisation and Embryology Act (HFEA) 2008 (c22). http://www.legislation.gov.uk/ukpga/2008/22/contents
19. Gilbert F, O'Connell CD, Mladenovska T et al (2018) Print me an organ? Ethical and regulatory issues emerging from 3D bioprinting in medicine. Sci Eng Ethics 24(1):73–91. https://doi.org/10.1007/s11948-017-9874-6
20. Vijayavenkataraman S, Lu WF, Fuh JYH (2016) 3D bioprinting—an ethical, legal and social aspects (ELSA) framework. Bioprinting 1–2:11–16
21. Enoch S, Shaaban H, Dunn KW (2005) Informed consent should be obtained from patients to use products (skin substitutes) and dressing containing biological material. J Med Ethics 31(1):2–6
22. European Parliament and Council of the European Union European Tissues and Cells Directive, Directive 2004/23/EC. https://eur-lex.europa.eu/legal-content/EN/TXT/?uri=celex:32004L0023
23. Human Tissue Act 2004, c. 30. https://www.legislation.gov.uk/ukpga/2004/30/contents
24. Varkey M, Atala A (2015) Organ bioprinting: a closer look at ethics and policies. Wake Forest Journal of Law and Policy 275–298

25. Harbaugh JT (2015) Do you own your 3D bioprinted body? Analyzing property issues at the intersection of digital information and biology. Am J Law Med 41:167–189

26. National Institutes of Health (NIH) (2009) Guidelines on Human Stem Cell Research. https://stemcells.nih.gov/policy/2009-guidelines.htm

27. Kantor J (2017) Public support in the U.S. for human-animal chimera research: results of a representative cross-sectional survey of 1,058 adults. Stem Cells Transl Med 6:1442–1444

28. Koplin JJ, Savulescu J (2019) Time to re-think the law on part-human chimeras. J Law Biosci 6:37. https://doi.org/10.1093/jlb/lsz005

29. Moy A (2017) Why the moratorium on human-animal chimera research should not be lifted. Linacre Quarterly 84:226–231

30. World Health Organization, Laboratory Biosafety Manual 3rd Ed. https://www.who.int/csr/resources/publications/biosafety/WHO_CDS_CSR_LYO_2004_11/en/

31. Bishop ES, Mostafe S, Pakvasa M et al (2017) 3-D bioprinting technologies in tissue engineering and regenerative medicine: Current and future trends. Genes Dis 4:185–195

32. Mandrycky C, Wang Z, Kim K et al (2016) 3D bioprinting for engineering complex tissues. Biotechnol Adv 34(4):422–434

33. Office of the Federal Registrar, Code of Federal Regulations Title 21, Parts 1270 and 1271. US Government. https://www.ecfr.gov/cgi-bin/text-idx?SID=3ee286332416f26a91d9e6d786a604ab&mc=true&tpl=/ecfrbrowse/Title21/21tab_02.tpl

34. World Health Organization (1997) Handbook: good laboratory practice (GLP): quality practices for regulated non-clinical research and development–2nd ed. https://www.who.int/tdr/publications/training-guidelinepublications/good-laboratory-practice-handbook/en/

35. Office of the Federal Registrar Code of Federal Regulations, Title 21, Part 58. US government. https://www.ecfr.gov/cgi-bin/text-idx?SID=59467274f688447a894544906d008a39&mc=true&tpl=/ecfrbrowse/Title21/21tab_02.tpl

36. OECD (1997) Principles of Good Laboratory Practice. http://www.oecd.org/officialdocuments/publicdisplaydocumentpdf/?cote=env/mc/chem(98)17&doclanguage=en

37. World Health Organization, Good Manufacturing Practices for Biological Products, Technical Report Series no 999, Annex 2. https://apps.who.int/medicinedocs/documents/s22400en/s22400en.pdf. Accessed 9 Aug 2019

38. Office of the Federal Registrar (2011) Code of Federal Regulations, Title 21, Part 210 and 211. US Government

39. European Parliament and Council of the European Union, Commission directive 2003/94/EC of 8 October 2003 laying down the principles and guidelines of good manufacturing practice in respect of medicinal products for human use and investigational medicinal products for human use. https://ec.europa.eu/health/sites/health/files/files/eudralex/vol-1/dir_2003_94/dir_2003_94_en.pdf. Accessed 9 Aug 2019

40. National Health and Medical Research Council NHRMC (2013) Australian code for the care and use of animals in research. Government of Australia, Canberra. https://nhmrc.gov.au/about-us/publications/australian-code-care-and-use-animals-scientific-purposes

41. Festing S, Wilkinson R (2007) The ethics of animal research. Talking Point on the use of animals in scientific research. EMBO Rep 8 (6):526–530. https://www.ncbi.nlm.nih.gov/pmc/articles/PMC2002542/pdf/7400993.pdf

42. The Animals (Scientific Procedures) Act 1986. https://www.legislation.gov.uk/ukpga/1986/14/contents

43. Animal Welfare Act of 1966 (AWA). https://www.gpo.gov/fdsys/pkg/USCODE-2015-title7/html/USCODE-2015-title7-chap54.htm

44. Animal Welfare Regulations. https://www.gpo.gov/fdsys/pkg/CFR-2016-title9-vol1/xml/CFR-2016-title9-vol1-chapI-subchapA.xml

45. European Parliament and Council of the European Union, Directive 210/63/EU on the protection of animals used for scientific purposes. https://eur-lex.europa.eu/legal-content/EN/TXT/?uri=CELEX:32010L0063

46. Ogden BE, Wanyong P, Agui T et al (2016) Laboratory animal laws, regulations, guidelines and standards in China Mainland, Japan, and Korea. ILAR J 57(3):301–311. https://doi.org/10.1093/ilar/ilw018

47. European Parliament and Council of the European Union, Regulation (EU) 2017/745 of the European Parliament and of the Council of 5 April 2017 on medical devices and Regulation (EU) 2017/746 of the European Parliament and of the Council of 5 April 2017 on in vitro diagnostic medical devices. https://ec.europa.eu/growth/sectors/medical-devices_en

48. Hourd P et al (2015) 3D-bioprinting exemplar of the consequences of the regulatory requirement on customized process. Regen Med 10 (7):863–883

49. Commonwealth of Australia (2018) Therapeutic Goods (Medical Devices) Regulations 2002. https://www.legislation.gov.au/Series/F2002B00237

50. Commonwealth of Australia (July 2019) Commonwealth of Australia (July 2019) Therapeutic Goods Regulations (1990). https://www.legislation.gov.au/Series/F1996B00406

51. Commonwealth of Australia (2011) Australian Regulatory Guidelines for Medical Devices. https://www.tga.gov.au/sites/default/files/devices-argmd-01.pdf Accessed 9 Aug 2019

52. Commonwealth of Australia Australian Regulatory Guidelines for Biologicals. https://www.tga.gov.au/publication/australian-regulatory-guidelines-biologicals-argb

53. European Parliament and Council of the European Union. Therapy Medicinal Products (ATMP) Regulation. Regulation (EC) No 1394/2007 of the European Parliament and the Council of 13 November 2007 on advanced therapy medicinal products and amending Directive 2001/83/EC and Regulation (EC) No 726/2004. https://ec.europa.eu/health//sites/health/files/files/eudralex/vol-1/reg_2007_1394/reg_2007_1394_en.pdf. Accessed 9 Aug 2019

54. Li P (2014) 3D bioprinting technologies: patents, innovation and access. Law Innov Technol 6:282–304

55. Bauer HKM, Heller M, Fink M et al (2016) Social and legal frame conditions for 3d (and) bioprinting in medicine. Int J Comput Dentist 19(4):293–299

56. Mendis D, Nielsen J, Nicol D et al (2017) The coexistence of copyright and patent laws to protect innovation: a case study of 3D printing in UK and Australian Law. The Oxford Handbook of Law, Regulation and Technology. Edited by R Brownsword, E Scotford, and K Yeung. doi: https://doi.org/10.1093/oxfordhb/9780199680832.013.80

57. Tran JL (2015) To bioprint or not to bioprint. N C J Law Technol 17:123–178

58. Nielsen J, Nicol D (2019) The reform challenge: Australian patent law and the emergence of 3D printing. 3D printing and beyond: intellectual property and regulation. Edited by D Mendis, M Lemley and M Rimmer. Edward Elgar Publishing, Cheltenham, UK, pp 325–346

59. European Patent Convention (EPC), Article 53(a). https://www.epo.org/law-practice/legal-texts/html/epc/2016/e/ar53.html

60. India, Patents Act, 1970, Chapter II, Section 3b (amendments). http://www.ipindia.nic.in/writereaddata/Portal/IPOAct/1_31_1_patent-act-1970-11march2015.pdf. Accessed 9 Aug 2019

Chapter 5

Extrusion-Based Bioprinting: Current Standards and Relevancy for Human-Sized Tissue Fabrication

Kelsey Willson, Dongxu Ke, Carlos Kengla, Anthony Atala, and Sean V. Murphy

Abstract

The field of bioengineering has long pursued the goal of fabricating large-scale tissue constructs for use both in vitro and in vivo. Recent technological advances have indicated that bioprinting will be a key technique in manufacturing these specimens. This chapter aims to provide an overview of what has been achieved to date through the use of microextrusion bioprinters and what major challenges still impede progress. Microextrusion printer configurations will be addressed along with critical design characteristics including nozzle specifications and bioink modifications. Significant challenges within the field with regard to achieving long-term cell viability and vascularization, and current research that shows promise in mitigating these challenges in the near future are discussed. While microextrusion is a broad field with many applications, this chapter aims to provide an overview of the field with a focus on its applications toward human-sized tissue constructs.

Key words Microextrusion printing, Bioprinting, Cell printing, Nozzle, Bioink, 3D printing, Vascularization, Rheology

1 Introduction

Extrusion printing is a form of 3D fabrication wherein one or more substances are forced out of a die orifice to form a long strand which can be laid down in specific patterns. Height is added to the design by stacking multiple layers of these filaments. Microextrusion refers to cases where the extrusion orifice has a diameter of less than 1 mm. Unlike other types of rapid manufacturing such as stereolithography or selective laser sintering, extrusion printing has the capability to quickly and accurately place multiple materials within a single construct. Laser induced forward transfer (LIFT) printing and droplet-on-demand printing can also create complex patterns with multiple cell types but extrusion printing can outpace them, applying material much more rapidly. This allows for larger constructs to be created. While microextrusion printing has

Jeremy M. Crook (ed.), *3D Bioprinting: Principles and Protocols*, Methods in Molecular Biology, vol. 2140,
https://doi.org/10.1007/978-1-0716-0520-2_5, © Springer Science+Business Media, LLC, part of Springer Nature 2020

strengths and weaknesses which will be discussed further in this chapter, its contributions to the field of bioengineering are undeniable. To date, microextrusion printing has been used to create structures ranging from in vitro disease models to cardiovascular tissue structures, bone scaffolds, and constructs for in vivo implantation.

The application of 3D printing to biomedical engineering has opened the door to a new era of development. The number of publications in PubMed with the keyword "bioprinting" increased from a single return in 2005 to 269 in 2017. Those marked with the keyword "3D bioprinting" have also increased exponentially, from 1 publication in 2007 to 200 publications in 2017. Bioprinting has been well received as an alternative to conventional manufacturing techniques due to its ability to fully control the shape and composition of the printed components [1]. Freeform printers, including microextrusion printers can create complex shapes with ease. In the field of biomedical engineering, researchers often use scans from medical imaging systems (magnetic resonance imaging [MRI], computed tomography [CT]) in combination with computer-aided design (CAD) software to create biologically relevant designs. This combination of imaging and complex fabrication allows for the creation of unique, individualized prints, which can be adjusted for each print iteration. This has strong advantages for personalized medicine, allowing for researchers to fit each design to specific patients and applications. Combining CAD with microextrusion printing has allowed for custom surgical models, and will 1 day allow for customized tissue replacements.

While microextrusion printers have many advantages and hold promise for creating clinically relevant tissues in the future, there is still significant room for improvement. Their ability to print multiple materials is countered by the fact that each material must be carefully tuned, and printing parameters must be determined including appropriate nozzle size/shape, printing pressure/temperature and the speed of the extrusion head. While materials not containing cells can be altered for different mechanical properties, when cells are introduced into the mixture researchers must consider how these modifications will affect cell viability. These can manifest through increased printing pressures and higher shear stresses both of which are known to decrease cell viability [2–4]. Damaging the cells during production can lead not only to significant cell death but also unexpected changes in cell phenotypes and functionality after printing [4–8]. Furthermore, cell viability is not solely dependent on the shear stresses created by the combination of nozzle and bioink. Using microextrusion to print large or complex constructs can take hours to complete and cells have a finite life span when exposed to printing conditions. Overly large or complex designs and their correspondingly increased print times may surpass this life span resulting in a construct with limited cell viability that is unsuitable for use in vitro or in vivo [2]. Oxygen and nutrient transport through large structures is also currently limited by a lack of fabricable vasculature, which reduces long term cell viability within the device.

The ability to manufacture full scale, functional human tissue should not be thought of as trivial. Despite huge advances in additive manufacturing, there are still many challenges to overcome. This chapter will provide an introduction to the role bioprinting continues to play in pursuing the creation of clinically relevant tissues by discussing key components of a microextrusion printer, how they each impact the final print, and some of the current limitations present in the field. Major challenges posed for creating functional human scale tissues will be reviewed along with techniques being developed to address these challenges and hypothesize what future progress may still be needed to achieve fully functional tissues via bioprinting. Primary among these topics will be factors that influence printability—how well the material is extruded, spatial control, and how well the final print conforms to the initial design—and those factors that impact cell viability.

2   Microextrusion Printers

Microextrusion printers have a number of essential components; a print bed, a set of moving axes, an extrusion orifice, extrusion mechanism, and a media reservoir. With these five components one can print cells. There are however, a variety of options for each of these individual pieces, as well as additional mechanisms that can be added to the basic setup. This section will touch upon alterations to each of the aforementioned pieces and will briefly canvass a few of the many modifications that can be made to your printer.

Every printer, whether 2D or 3D, comes with a print bed. This is the surface on which your substrate is applied. In the case of conventional inkjet printers this is a piece of paper, while laser and lithography can print directly into a vat of liquid media. For many extrusion 3D printers the print bed is a flat piece of plastic, glass, or even silicone. When printing with cells, it is important to choose a material that will allow the cells to continue growing after printing is completed. Therefore, petri dishes or glass slides are often selected, though there are systems that print into vats of supporting media, and printers are being developed which could print directly within the body without requiring an external build plate [9, 10].

Every printer has a media reservoir and extrusion orifice. In microextrusion bioprinting, the extrusion orifice is more typically referred to as the nozzle. Choosing an appropriate nozzle has a major impact on the final resolution of a print, how a specific material will extrude, and what shear stresses will be applied to the material during the printing process. This selection is vital to obtaining a satisfactory print, and will be discussed in detail later in this chapter. The media reservoir is where a bioink is held prior to being extruded through the nozzle. In extrusion printers these are syringes with the nozzle attached to the end. These syringes are typically metal or plastic. In addition to modifying the material the syringe is made of, systems today enable independent control of the environment for each syringe. Heating and cooling apparatuses are sold which maintain a syringe and its contents at a specific temperature. This is vital for ensuring materials are held at an extrudable viscosity and can be used to maintain optimal conditions for cell viability. Many bioprinters currently available allow users to incorporate multiple materials in their prints by designing the printer to hold a series of syringes. By alternating the head actively being used to print, researchers can create heterogeneous prints containing multiple bioinks. The separation of materials into their own syringes with independent nozzles allows each to be tuned for the specific bioink.

There are two main techniques used to physically extrude materials from a syringe through a printing nozzle, pneumatic and mechanical. Pneumatic printers use air pressure to extrude bioinks, while mechanical printers use either a piston or screw to physically extrude material as seen in Fig. 1. Both of these methodologies result in continuous pressure applied to the system allowing filaments to be extruded; but each has independent strengths and weaknesses.

Pneumatic printers use a compressed gas to force bioink from a syringe during printing. These setups have relatively simple components that allow them to dispense materials over a wide range of pressures [12]. The ability to use high pressures with these systems allows researchers to print with highly viscous materials. This

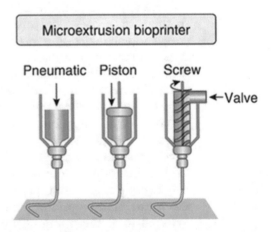

Fig. 1 Microextrusion printheads showing extrusion via pneumatic, piston, and screw dispension [11]

includes not only bioinks, but also molten polymers that can be used as structural support in bioprinted designs. Since the pressure can be easily manipulated, a single printer can be calibrated for a large number of materials using straightforward adjustments to pressure and printing nozzle [1, 13, 14]. However, the highly variable pressure must be carefully regulated with bioinks that contain cells to maximize cell viability after printing, as higher pressures can result in greater shear stresses at the nozzle outlet leading to cell death [2–4]. Additionally, the use of compressed gas generally means that there is as small delay in dispensing which can contribute to lower resolution.

Mechanical systems use either a piston or screw to dispense bioink. These designs do not exhibit the extrusion delay found in pneumatic printers. Piston printers in particular have increased control over the physical deposition of bioinks due to direct mechanical displacement [15]. However, this methodology is limited in the amount of pressure it can provide, thus making it unsuitable for certain high viscosity inks. Screw designs have been shown to be excellent for dispensing high viscosity fluids with finer spatial control [16, 17]. Due to the unique spatial configuration of screw driven dispensers, they are susceptible to increased cell death due to larger pressure differentials at the nozzle [17, 18]. While all three types of dispensing mechanisms (pneumatic, screw, and piston) have been used successfully in 3D printing and can be used in bioprinters it is important to consider their strengths and weaknesses when selecting the appropriate dispenser, and to carefully adjust each for optimal performance when printing with biological components.

The final component, one that that truly makes a 3D printer capable of 3D fabrication, is the moving axes. There are several different types of axes setup utilized in both handmade and commercial bioprinters. The most typical is a three-axis system that has stepper motors allowing motion independently in the x, y, and z directions, detailed in Fig. 2. This allows the print head to move freely in the xy plane, while the z-axis can be adjusted for specific layer heights and can be accomplished in two ways. The first is by either moving the print head, the build plate, or a combination of the two components (e.g., moving the building plate in the xy plane while the nozzle moves along the z-axis). Printers with movement along each axis independently, are known as Cartesian printers and are the most common type of printer used in extrusion based bioprinting. Similar motions can be replicated by delta printers that have a nozzle suspended over a print bed, controlled by a three guide arms (diagrams of both arrangements are shown in Fig. 2). The nozzle location for these systems is dictated by the height and angle of each of these guide arms, while the print bed (and final construct) remains stationary. These printers have limitations with regard to printing free-standing features including

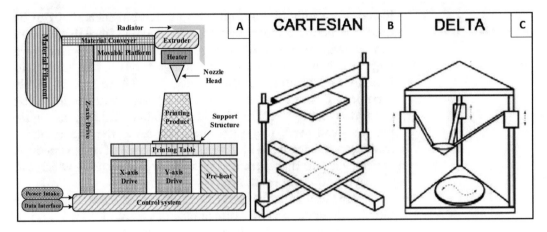

Fig. 2 Detailed schematic of a Cartesian printer (**a**) and diagrams of Cartesian and delta movements (**b, c**) [19, 20]. (**b**) and (**c**) were adapted from Kleefoot under CC4.0 [21]

overhangs or hollow components and the layer-by-layer approach along the z-axis can result in reduced mechanical integrity for the final construct, but are the most commonly used systems for micro-extrusion bioprinting.

Despite their common usage, the limitations seen with Cartesian printers have driven researchers to develop printers with an increased number of axes. Several groups have debuted printers with five axes, wherein tilt and rotation can be controlled along with translation in the *xyz* coordinates shown in Fig. 3. This allows researchers to maneuver the sample to eliminate distinct layers at positions of high strain and to print severe overhangs, creating stronger, lighter constructs [19, 20]. Printers have also been developed with six axes, either by combining two three-axis systems (one applied to the print head, the other to the print bed) or by using six-axis robotic arms as seen in Fig. 3 [21–23, 25]. These, similarly to five-axis printers, can be used to fabricate complex overhangs and alter the orientation of layers within a printed construct. While these systems are still in a nascent stage of development it is important to note that six-axis printing is already being applied to bioprinting [26]. As further improvements are made to standard 3D printers, they can be quickly incorporated into the field of bioprinting allowing researchers to continue making more complex, and biologically relevant, structures.

3   Commercial Bioprinters

As 3D printing has moved into the public view, companies such as Organovo, Allevi, RegenHu, Cellink™, Advanced Solutions, and many others have emerged in the commercial field. These companies have taken research that conventionally has only been available

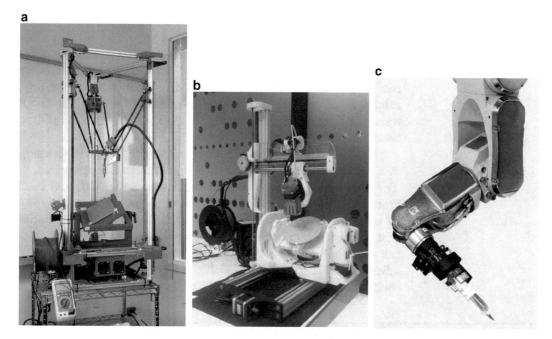

Fig. 3 Examples of printers with increased axes. A five-axis delta printer (**a**), a five-axis Cartesian printer (**b**), and a six-axis robotic arm (**c**, Image: Advanced Solutions Life Science) [22–24]

to those able to build their own systems and made bioprinting accessible to independent companies, academic labs, and large corporations. The technology is currently being used for a wide variety of applications including in vitro tissue testing, personalized medicine and the development of new medical devices. Importantly, in the quest for human scale tissues these companies are developing building blocks: bioinks, printing processes and new printers, giving a wider range of groups the ability to test innovative ideas and theories in their own labs.

There are multiple bioprinters on the market currently, ranging from the INKREDIBLE and RegenHu printers which are micro-extrusion based [27, 28], to more exotic systems including the Regenova printer by Cyfuse which fixes cell spheroids in place on a microneedle array before merging the cells into a single construct [24] or the BioassemblyBot® by Advanced Solutions Life Sciences which uses a six-axis robotic arm to precisely control the location of extruded material [26]. These printers have been used to develop a variety of tissues in vitro including cartilage, liver, bone, and trachea [27–34]. While researchers still struggle to produce functional human-scaled tissues, the development of in vitro models has proved invaluable for testing new drug therapies and studying disease development [32, 35–39].

With every advance made using in vitro studies, the field comes one step closer to human scale tissues. This extends beyond the immediate printing of tissues and into foundational research on creating new bioinks, new printing protocols, and new methodologies to keep tissue functional both in vitro and in vivo. The standardization of bioinks as companies develop and characterize new substrates both with and without cells [40–43] will allow for more reproducible prints done at differing locations. Similarly, standardized printing protocols/techniques made public to all researchers assists in streamlining the path to full scale bioprinted tissues.

Commercialized printers have expanded the field of bioprinting to a larger community of scientists, researchers and entrepreneurs. They have been used to create tubular structures, soft and hard tissues, drug testing devices, and disease models in vitro [27, 44–47]. While still limited by scale, speed, and resolution, difficulties that plague the bioprinting field as a whole, commercial bioprinters can print a wide variety of structures and researchers are taking advantage of these readily available systems to speed the development of fully functional human tissues.

4   Key Factors for Extrusion Bioprinting

4.1   Bioprinter Setup: Selecting the Right Nozzle

While there are a multitude of printer configurations, as previously discussed, the print nozzle is the one component that is universal to all systems, be it homemade or commercial, piston or pneumatic. This small, ubiquitous attachment is one of the most crucial components in setting up your bioprinter. The nozzle selection will have significant effects on the shear stress, bioink distribution, and rheological properties of the extruded bioink. In the following section, we will discuss choosing a nozzle based on its shape and size, and what extra functions researchers have incorporated into the dispensing orifice. It should also be noted that new nozzles are constantly being developed to further expand the applications of bioprinting. While this section focuses on the common parameters of size and shape, additional functionality provided by modified nozzles for coaxial printing will be discussed with regard to creating vasculature within engineered constructs. Further, though the nozzle plays a critical role in bioprinting, many of the other parameters, including pressure and bioink feed rate, are dependent on the rheological properties of the bioink and must be optimized after the nozzle is selected to obtain high shape fidelity and printing resolution while maintaining high cell viability.

4.1.1   Nozzle Shape

The first criteria for selecting a nozzle is the nozzle shape, which directly affects the shape of the bioink when it is extruded. Besides controlling the shape of the filament, the shape of the nozzle plays a crucial role in the shear stresses applied to the bioink, which have significant effects on cell viability during printing [14, 18]. Billiet et al. reported a study for the preparation of a bioprinted cell-laden tissue engineered constructs using gelatin methacrylamide hydrogel [14]. In this study, the authors discussed the effects of the nozzle shape (cylindrical vs. conical) on cell viability. Using finite element simulations, they were able to correlate experimental data to a model for shear stress distribution as a function of nozzle shape. The simulated model showed that conical nozzles are preferable over cylindrical nozzles at low printing pressures ($<$300 kPa) in order to maintain high cell viability. However, the model also showed that these advantages diminished when the printing pressure was increased.

4.1.2 *Nozzle Size*

The nozzle size must be considered in regard to both the nozzle diameter and length. Both influence shear stresses experienced by the substrate during extrusion and therefore affect cell viability. Chang et al. reported a study discussing the effects of nozzle diameter on cell survival rate using a direct cell printing method [48]. The results showed that increasing the nozzle diameter could lead to higher shear forces which reduced the cell survival rate. When the nozzle length was increased, the time in which the cells were held in the printhead increased, resulting in reduced cellular viability [49].

Furthermore, nozzle diameter has a direct impact on the resolution of microextrusion printers. This is a major limitation, particularly when compared to deposition techniques such as LIFT, which can deposit single cells. While many extrusion systems boast resolutions as fine as 1 μm, this generally refers to the ability to precisely place the extrusion needle. The actual printing resolution is dependent on a number of factors, including the material viscosity, the flow rate of the material and the needle diameter. It is generally accepted that when printing with cells, extrusion needles should be kept at 150 μm and larger to minimize needle clogging and high shear stresses while the cells are extruded [1, 48]. This limits the actual resolution of bioprinters significantly more than the reported positioning resolution [8]. In the context of bioprinting, resolution is determined by the diameter of the extruded filament, for inks with and without cells, and the accurate placement of the extruded filaments. Selecting appropriate nozzle dimensions will significantly affect this parameter, and as such, is not only vital for ensuring cellular viability but also a critical component of designing and executing high resolution prints.

4.2 Bioink Rheology

The printability of a material is generally controlled by the rheological properties of the material and must be adjusted in order create constructs with long-term high shape fidelity. Each different material has unique characteristic rheological properties. Some materials, such as collagen [50, 51], cellulose [52], and gelatin [53], can be obtained from a variety of natural sources and have appropriate rheological properties for bioprinting. Other materials, such as alginate [54] and hyaluronate [55], are commercially available and widely used in the bioprinting, despite suboptimal printing properties. In most cases, these materials are combined together to achieve improved printability along with other beneficial properties for tissue engineering applications. Markstedt et al. reported a novel bioprinting study using nanofibrillated cellulose (NFC) and alginate combined with human chondrocytes, used to make tissue engineered ears [56]. The pure alginate used in this study had a poor shape fidelity during the printing, however, with the presence of 2.5% NFC, the bioink presented excellent rheological properties with improved shape fidelity and high printing resolution. Vanderhooft et al. also reported a hydrogel toolkit using hyaluronate and gelatin with a wide range of rheological properties [57]. The gels in this kit could be combined in differing concentrations, and these different formulae results in storage shear moduli ranging from 11 to 3500 Pa, which was beneficial for different tissue engineering applications. Further work has been completed in other laboratories to further the use of similar hydrogels in bioprinting. Skardal et al. showed that when both hyaluronate and gelatin were methacrylated, the hydrogel gained the ability to be photocrosslinked, which further improved its rheology properties and extended its applications in the field [15].

Even though there are many materials that are suitable for bioprinting, scientists are still trying to develop new bioinks to improve the properties of final constructs. When new materials are developed, a two-step procedure is commonly used to assess how amenable they are with bioprinting [58]. The first step is manual dispensing to see if this new material can be extruded to form distinct fibers and if these fibers can be stacked to build a 3D construct. The second step is to test the rheological properties using viscosity and recovery tests which provide further information about the bioink and may provide insights toward modifications that can improve its printability. Once the candidate material passes through these tests, other steps will be needed to improve the mechanical stability of final constructs. The development of gelatin methacrylate (GelMA) is a perfect example of a material being modified to pass these tests, resulting in a material suited for bioprinting. Gelatin itself is a printable substrate; however, with the chemical modification, GelMA can be covalently cross-linked under UV to form stronger constructs compared to gelatin [15, 59].

5   Bioprinting Challenges

Bioprinting has enabled significant advances in the field of bioengineering, but as with any technology it has limitations and challenges. For those who have bioprinters at their disposal, day to day challenges include fine-tuning the setup for different inks, picking the right nozzle, and generally fine tuning the key printing factors discussed above. As a field however, bioprinting faces a much larger set of challenges if it is to create human scale-tissues. Here, three challenges critical to the development of large-scale constructs will be discussed in detail: choosing or creating appropriate bioinks to ensure appropriate mechanical characteristics, creating vascularized prints, and ensuring cellular viability before, during and after the printing process. While there are many groups working on these challenges, each one still acts as a major hurdle in the development of clinically relevant bioprinting. This section will cover the importance of each of these challenges, the current state of the field, and how researchers are working toward better solutions.

5.1 Bioinks

The ability for extrusion bioprinters to accommodate many different types of bioinks enables them to print complex shapes and high cell densities. These bioinks are the foundation of each printed construct. Careful selection of the initial bioink impacts how the construct can be printed, the final cell viability, and mechanical properties of the final device. Many microextrusion printers capitalize on the differences between bioinks by printing different materials into the same device, combining mechanically, physically, and chemically unique bioinks into scaffolds suitable to different environments and cell types. This allows researchers to tune inherent mechanical properties of bioprinted tissues and organs, matching those seen in native tissues to achieve similar functionality both in vivo and in vitro.

Industrial extrusion printers have typically used plastics as their extrusion media. Bioprinters, adapted from these conventional systems, have retained the ability to process structurally stable materials, including polymers like polycaprolactone (PCL) which have been used both as scaffolds for cell growth and to provide mechanical support for other materials in the same construct. Extrusion printers have also been used to print a wide variety of hydrogels. These gels, while generally soft due to a naturally high water content, can be gelled or reinforced to create functional scaffolds [13]. Hydrogels, while mechanically weaker than conventional 3D printing substrates, allow bioprinters to incorporate cells into the construct. As researchers have moved forward with different cell lines in various labs, a wide selection of compatible hydrogels has been developed as bioinks. With the multitude of options now available, and new inks being constantly developed and improved upon it is important to understand how each can be tailored to fit the needs of a specific print job. With such a large selection of printable materials this section will not attempt detail individual bioink recipes but rather will introduce and discuss generalizable approaches used to tailor the mechanical properties of bioprinted constructs that will be applicable to many bioinks.

The mechanical properties of bioprinted constructs are directly dependent on the chemical composition of the bioink. However, it is not always possible to find materials that already fulfill the mechanical requirements of every tissue type. In these situations, chemical modifications can be made to tailor the mechanical properties of a material to get it closer to the final goal. Hyaluronic acid (HA) is a widely used material that presents in the extracellular matrix and provides mechanical support. Despite this, pure HA is too soft for some tissue engineering applications, such as bone and cartilage. It is also prone to undergo fairly rapid degradation, which is not always desired as it may result in a lack of structural support before the cells/tissue can build their own matrixes. Methacrylated hyaluronic acid (MeHA) is a chemically modified HA which can be further cross-linked by the addition of a photoinitiator under UV exposure. This results in a network formation. It has been reported that the compressive modules of gels created using MeHA, could be tailored from 2 to over 100 kPa by varying the molecular weight of the initial HA from 50 to 1100 kDa and changing the degree of methacrylation from 2 wt.% to 20 wt.% [60]. Poldervaart et al. reported using MeHA as a bioink when creating constructs for bone tissue engineering applications [61]. The MeHA bioink showed excellent mechanical properties and printability. In addition, the embedded human bone marrow derived mesenchymal stromal cells (MSCs) showed a cell viability of 64.4% after 21 days of culture and proper osteogenic differentiation when in the presence of bone morphogenetic protein-2 (BMP-2).

Pure gelatin is a water-soluble protein that has thermoreversible properties which allow it to form hydrogels [62]. It plays a crucial role in bioprinting 3D models and scaffolds for tissue engineering applications, however, it is not stable at body temperature. Further, it has proven difficult to tailor the mechanical properties of gelatin for different applications. By chemically modifying gelatin with unsaturated methacrylamide groups, a cross-linkable polymer, gelatin-methacrylamide (GelMA), is created that can then be tailored for various mechanical properties and which can encapsulate cells and drugs for tissue engineering and target drug delivery applications [59, 63–69]. Chen et al. reported using GelMA as the primary bioink to print human blood-derived endothelial colony-forming cells (ECFCs) and MSCs, which were able to form a vascular network [69]. Increasing the degree of methacrylation has been shown to improve the compressive modulus of GelMA, indicating the possibility that GelMA may be used for a variety of tissue engineering applications including as a bioink for the preparation of 3D printed cartilage constructs [59]. By increasing the concentration of GelMA and the UV exposure time, the compressive modulus was increased while the percentage of effective swelling decreased. Equations were also reported based on the results that allowed the researchers to calculate best-fit parameters for models with different compressive moduli. In addition, this bioink was able to maintain high viability for cells that were encapsulated after 3 days of culture and was able to induce glycosaminoglycan and collagen type II formation after 4 weeks of culture.

5.1.2 *Hybrid Bioinks*   Another way to improve the mechanical properties of bioprinted constructs is through the use of hybrid bioinks. These hybrids can be designed to make up for the disadvantages of each individual material. In addition, the individual components within the hybrid bioink can be chosen to ensure the potential for interconnected cross-linking, which allows for further improvement and control of mechanical properties for specific tissue engineering applications. HA and collagen are major components of the extracellular matrix (ECM) found in human tissues. Gelatin is an irreversibly hydrolyzed form of collagen and a commonly used hydrogel for bioprinting. Hence, the hybrid bioink containing HA and gelatin can be used to mimic ECM and its role in the mechanical support of native human tissues. Camci-Unal et al. reported a cross-linkable hybrid hydrogel containing GelMA and MeHA with mechanical properties that could be tailored for different tissue engineering applications [70]. The increase of GelMA and MeHA concentrations improved the compressive modulus of the hybrid hydrogel. In addition, the presence of gelatin methacrylate (GelMA) in MeHA induced the cell spreading in the hybrid hydrogel compared to the pure MeHA hydrogel. Another study reported a simple toolkit to prepare hydrogel bioink for fulfilling the mechanical requirement of different tissues [71]. By changing the cross-linkers and their concentrations, the storage modulus could be shifted from 100 Pa to 20 kPa. This range is applicable to the preparation of liver, heart, skeletal muscle, bone marrow, fat, brain/nerve, lung, kidney, and smooth muscle tissues. Further, the group included information on creating mechanical gradients for complex tissues and tissue connection sites, which has proven difficult to achieve. A study reported by Merceron et al. demonstrated the manufacturing of a muscle–tendon unit including polyurethane (PU)/C2C12 side for elasticity and muscle development and poly(ε-caprolactone) (PCL)/NIH/3T3 side for stiffness and tendon development [72]. The PU side was elastic and had a tensile modulus of 0.39 ± 0.05 MPa, while the PCL side was stiff and had a tensile modulus of 46.67 ± 2.67 MPa. In addition, the bioprinted muscle-tendon unit maintained a high cell viability on both sides after 7 days of culture and supported cell differentiation toward myoblasts on the muscle side and fibroblasts on the tendon side.

5.2 Vascularization   Arguably one of the largest challenges facing the bioprinting community today is creating a vascularized construct. Unlike the development of new bioinks, which relies on changing components used to print with, creating a vascular network within an engineered construct requires changes in printing techniques and, in some cases, the hardware used to print with. The addition of vasculature to 3D printed constructs is vital as vascularization is key for oxygen, nutrient, and waste transport. Within a tissue, oxygen is consumed by cells faster than it can diffuse. This results in an oxygen deficiency as distance from a vessel (oxygen conduit) is increased. Practically speaking, this means that relatively few cells can survive if they are more than 200 μm away from a blood vessel [73–75]. Nutrients are absorbed in a similar manner. Limitations on oxygen and nutrient access restrict the size of biological constructs that can be created and maintained by scientists. The development of a robust methodology to address the addition of vascularization in manufactured tissue constructs would allow for the development of large tissues, potentially full-scale organs that could be translated toward clinical use.

Vascular systems are not relied on solely for delivery of needed compounds like oxygen and nutrients. It is also necessary for the removal of waste products that accumulate in tissue. Without proper vascularization to assist in removal, waste products can build up over time, creating cytotoxic environments [76]. This is of particular importance for many bioengineered applications, due to the use of scaffolds. As implanted scaffolds degrade, they break into a variety of waste components, many of which can be toxic to cells. Without vascularization throughout an implanted scaffold to remove waste products, this degradation can result in necrotic zones at the implant site, inhibiting tissue repair and regeneration.

Blood vessels are currently manufactured using a variety of techniques. These include scaffold processing, cell sheets, casting, and decellularization of natural vessels [77–82]. While these techniques are able to create blood vessels, they suffer from a number of restrictions. These methodologies are limited in the sizes and shapes they can create [83, 84]. Manufacturing processes can be complex and time consuming, which often results in low cost efficiency [77, 85]. In addition, vessels made using these techniques can have high rates of thrombogenicity along with the possibility for atherosclerotic degeneration and intimal hyperplasia [80, 82, 85].

While there has been limited success in the development of grafts for large artery replacements, currently there are few options for small diameter vessels (<6 mm diameter) [86]. When the desired diameter drops to capillary dimensions, there are even fewer fabrication techniques. When approached clinically, this may initially seem of minimal importance as vascular surgeons are not likely to need individual capillary replacements for patients.

However, as discussed previously, this disparity is of critical importance in the quest to develop fully functional tissue constructs. As such, there are many groups working on new approaches that address the inadequacies with current vascular manufacturing techniques. Here we will outline three promising 3D-fabrication strategies that are currently under development to solve these limitations. Each of these showcases the development of new techniques or innovative combinations of established methodologies, that when combined with microextrusion printing could lead to the incorporation of complete vascular networks in engineered tissues.

5.2.1 Sacrificial Materials

Previously, bioinks have been discussed with regard to their impact on cell viability, and how they contribute to the mechanical characteristics of the final construct. However, when selected appropriately, these inks can also be used as temporary components of a printed structure allowing for new configurations to be created. Materials used in this manner are known as sacrificial materials. They have a long history in conventional 3D printing, having been used to provide overhangs (unsupported/freestanding features in the design) and hollow internal spaces, but are now being applied for bioprinting as well [87].

Briefly, a sacrificial material refers to a substrate designed to be removed from the final product. Combining these removable pieces within bioink designs, which are often too soft to support hollow spaces during the printing process, has allowed researchers to create open lumens. Sacrificial materials have been used to create hollow vessels ranging in size from 800 μm down to nanotubes [88, 89]. Support structures can be incorporated into the final design by printing them independently then casting a cell-laden hydrogel around the sacrificial components. This has been successfully demonstrated in experiments using carbohydrate glass, which was printed as a lattice then surrounded with cell-laden hydrogels. These hydrogels were cross-linked before the carbohydrate glass was removed leaving a cellular construct interspersed with hollow lumens [88]. When cultured with primary hepatocytes, this method of adding vascular channels led to increased albumin and urea secretion with high cell viability after 8 days in culture [88]. This technique has also been used with cell-laden hydrogels to create thick perfusion chips. Pluronic, a compound with thermally sensitive gelation, was coprinted with cell laden bioinks. The entire printed matrix was encapsulated in a gelatin–fibrin solution. Once the system was cross-linked with thrombin, the pluronic is removed leaving a cellularized construct punctuated with hollow lumens [90].

Sacrificial materials can also be printed simultaneously with the surrounding substrates without casting. Polymers that undergo thermally reversible gelations are of particular interest for this application. One such polymer is pluronic, which has been used as a sacrificial material to create hollow lumens by exploiting its thermal gelation properties, specifically the fact that it is liquid at low temperatures and gels at body temperatures [91]. In practice, this has been used to print pluronic along with methacrylated gelatin creating 3D constructs which contain a network of pluronic filaments surrounded by cell-laden gels. Once the gels have been crosslinked, the constructs can be cooled briefly, liquefying the pluronic and leaving hollow vessels which can be perfused with media [92]. This technique has been used to create systems containing multiple cell types resulting in structures with controlled cell layouts and perfusable lumens [92]. It has been further expanded upon to create kidney's-on-a-chip and independent microvascular networks [93, 94].

Sacrificial materials can also be used as support materials by capitalizing on their buoyant forces. Buoyant forces have conventionally been used in drop-wise printing, as the aliquots are small enough to be supported on the top of a liquid purely due to buoyant forces [95, 96]. This has been shown to allow for the fabrication of severe overhangs and hollow/branching structures without deposited support material [97]. Recently this technique has been combined with microextrusion printing in the development of freeform reversible embedding of suspended hydrogels (FRESH) printing [10]. The FRESH method dispenses cell-laden biomaterials directly into a slurry of gelatin microparticles, which act as the buoyant/sacrificial material. This slurry is tuned so that it acts as a solid until a threshold shear force is applied, which allows the printed material to be supported while the extrusion head can still move through the remaining material. After printing and crosslinking, the system is warmed to liquefy and remove the gelatin leaving only the printed construct. A wide variety of biological systems have been replicated using this, including bifurcated tubes, miniaturized femurs and embryonic chick hearts [10].

5.2.2   Coaxial Bioprinting   The majority of microextrusion printers utilize standard single nozzles which can extrude rod-like lines which can be aligned to form complex patterns through layer-by-layer processing. However, these types of nozzles have difficulty forming certain shapes, such as blood vessels which require a hollow tubular structure. In the attempt to address this deficiency, several groups have reported on the use of coaxial printing to create small diameter tubules which have the potential to be used as vessels in engineered structures [54, 98]. First reported by Zhang et al., coaxial printing is an off-shoot of traditional microextrusion printing where two materials are extruded simultaneously [54]. This is done by using two needles for extrusion, one placed concentrically inside the other. One of the two printing materials is extruded through the center needle, and the other material is extruded in the space between the two needles. This results in a cylindrical "core" wrapped in a secondary layer. The printing speed and dispensing rates can be varied to create different diameters for both the inner and outer materials [99–101]. The materials that are printed can also be modified to allow for increased cell viability and adhesion [77, 98, 102]. To date, coaxial printing has achieved vessels ranging from 135:309 to 1800:2500 μm (inner diameter: outer diameter) with cells incorporated into the bioink [102, 103].

There are several materials that have been used to print with coaxial nozzles both with and without cells, and studies have shown that these prints result in perfusable, vascular-like constructs. Jia et al. reported the use of a coaxial nozzle to print a mixture of GelMA, sodium alginate, and four-arm poly(ethylene glycol)-tetra-acrylate (PEGTA) [98]. This mixture undergoes ionic cross-linking between sodium alginate, extruded through the outer nozzle, and calcium chloride, extruded through the inner nozzle, to create a temporarily stable structure during the print time. After printing, covalent photocrosslinking was performed through the reaction between GelMA and PEGTA under UV exposure, resulting in final vascular constructs that were permanently fixed and mechanically stable. This work has been expanded on, with Yu et al. reporting the use of coaxial nozzles to create tissue engineered cartilage-like tissues with tubular channels and hypothesizing on the future development of further nested needles such as triaxial nozzles [102]. Further, a recently published paper from Gao et al. reported the use of this technology with the first published fully endothelialized blood vessels created through direct printing [77]. Their research showed that engineered blood vessels, printed with epithelial cells and growth factors in the bioink, allowed for the development of arterioles and capillaries when used to replace damaged vessels in an ischemic model [77].

The use of coaxial printing provides the unique opportunity to create blood vessels through direct manufacturing rather than secondary or tertiary processing steps. Preliminary research has shown that this fabrication technique does not damage cells, but results in viable constructs which are robust in vitro and in vivo [54, 77, 102]. Though still a new technology, its ability to create multiple vessel diameters, including small diameter vessels, makes it an exciting development as researchers strive to solve the vascularization problem.

5.2.3 Self-Assembly

The majority of current vascular manufacturing approaches focus on fairly large vessels due to constraints with resolution. In recent years, self-assembly has shown promise in creating vessels on the scale of 5–200 μm, similar to the sizes seen physiologically for capillary and even microcapillary structures [75, 104]. Self-assembly is the spontaneous formation of structures and patterns from a mixture of component parts. This technique can be used to induce angiogenesis without prepatterning it by culturing endothelial cells with other cell types and allowing growth factors to drive the development of vessels. Microvessels have been created this way after coculturing endothelial cells with pericytes in fibrin for 5 days. The manufactured gels showed lumen densities ranging from 420–940 lumens/mm^2 and exhibited 40% perfusion after 6 days in vivo [104]. Self-assembly has also been used to create a vascular network around cellular spheroids in vitro. Endothelial cells were cocultured with fibroblasts in microfluidic channels on either side of a well containing a spheroid. Over the course of 1–3 weeks, angiogenic sprouts appeared reaching toward the suspended spheroid. These sprouts matured into vessels that showed no significant leakage and were able to support spheroids with a diameter of 1000 μm with minimal core necrosis [75]. Similar results have been reported when endothelial cells and fibroblasts were cocultured without a spheroid. The cells were able to reach across the separating gel to form an interconnected network within 5–7 days [105].

These approaches indicate that researchers may not be required to prepattern the smallest of capillary networks necessary to maintain a healthy tissue, but rather by providing the appropriate cell types and signals, microcapillary beds can be encouraged to develop on their own in vitro. Current 3D printers are limited in their resolution, with the general consensus being that needles should have a minimum diameter of 150 μm to reduce clogging and cell death [1, 48]. This places a limit on the smallest patternable features that can be achieved through 3D printing. By incorporating techniques such as self-assembly, researchers could theoretically print large vessels intermingled with parenchymal cells as demonstrated in the microfluidic chambers discussed above, to achieve a continuous network of vessels with a gradient of diameters mimicking that found in vivo.

5.3 Maintaining Cell Viability

While there are many different ways to make fabricate cellularized constructs, even within the field of bioprinting, extrusion based bioprinters have several key advantages over their droplet and laser deposition based counterparts. One of the most important when it comes to fabricating human scaled tissues is cell density. Extrusion bioprinters can print inks containing very high concentrations of cells which accelerates growth and the process of tissue formation [13]. Achieving a high cell density is critical to creating physiologically relevant tissues. Where other printing methodologies may have difficulty obtaining high cell density in a large construct, microextrusion printers can support bioinks with very high cell densities without significantly impacting the final print resolution or time to completion [106]. Several studies have shown that the cell concentration can also be increased in microextrusion bioinks without impacting cell viability allowing researchers to create constructs with very high cell densities using easy processing techniques [13, 18]. However, being able to increase the cell density while maintaining the same viability does not always guarantee that cell viability is high to begin with. While cells freshly harvested from confluent flasks may have measured viability nearing 90%, the bioprinting process and subsequent culturing conditions can be detrimental to cell survival. In this section we will discuss some of the parameters that directly affect cellular viability during and after the printing process.

Mechanical interference plays a vital role in cell viability during the printing process. During the extrusion, cells within the bioink usually experience shear forces, which can reduce cell viability. This shear force is directly related to the nozzle parameters, extrusion pressure, rheological properties, and cell density within the bioink. It has been reported that the increase of extrusion pressure and nozzle length as well as the decrease of nozzle diameter can result in higher shear forces, which significantly reduced the cell viability of HepG2 cells during printing [48, 49]. With regard to the bioink formulation, increasing the viscosity and decreasing the cell density also resulted in higher shear forces, which again reduced cell viability [49]. The printing speed effects cell viability in a different manner. When the printing speed is high, the extruded strands can be "stretched" leading to a high tensile force being placed on the embedded cells that can lead to reduced cell viability [48, 107]. Decreasing the printing speed can reduce this tensile force, while over compensating can lead to this tensile force being changed to a compressive force, which again will reduce cell viability. Finding an appropriate printing speed is tantamount to ensuring that you can print with live cells.

Cell viability after printing can be discussed as two distinct phases. During printing, the user needs to be cognizant of what is happening to cells that have been deposited into the construct as well as those still being extruded. Until the sample is complete, these cells must stay in the print zone on either plates or other substrates and are generally exposed to the local environment during this time. If this environment is not adjusted for maximum cell compatibility this exposure time can result in cell death. There are two major factors that researchers can adjust to minimize this drop in cell viability. First, increasing the printing speed can reduce this exposure time, allowing for the construct to be quickly transferred to an incubator. This modification is particularly crucial for large and complex constructs, which by their nature have longer print times. Secondly, researchers can tune the printing environment to mirror that found in a cell incubator. By incorporating high humidity and temperatures the cells will maintain a higher viability through the duration of each print.

Immediately following the completion of printing, cell viability can be as high as 90%.

The next step in maintaining cell viability is placing the construct in a culture chamber until it exhibits comparable functionality to natural tissue. This can be a time-consuming process that requires large volumes of nutrients and oxygen in order for the cells to proliferate and differentiate. In vivo, oxygen and nutrients are delivered via vasculature. As previously mention, this makes the design of vascular structure and improved vascularization within engineered tissue constructs crucial to facilitating cell growth and minimizing the tissue necrosis. Approaches for improving the vascularization in bioprinted constructs were reported in previous sections; however, with current manufacturing techniques, the rate is vascularization is still unable to provide adequate oxygen/nutrients to the cells in large constructs and the samples suffer decreased cell viability as a result. While there are a multitude of researchers working on ways to increase vasculature in printed constructs through manufacturing techniques, there are other methods of addressing this discrepancy. One approach is to reduce the cellular metabolism, creating a system that requires less oxygen and nutrients while maintaining high rates of cell proliferation and differentiation. Adenosine is a purine nucleoside which has been reported to play a crucial role in energy transfer, reducing neuronal energy requirements and decreasing neuronal excitability [108, 109]. The use of adenosine in the bioprinted construct should slow down the cellular metabolism giving researchers a longer window immediately following printing to establish a self-sustaining vascular network within bioprinted constructs. If the cellular metabolism cannot be adequately slowed, the incorporation of particulate oxygen generators (POGs) within the construct may be another useful approach in compensating for the lack of established vasculature. POGs should not interfere in the printing process, but once incorporated will provide an in situ supply of oxygen that should improve cell viability. The application of both adenosine and POGs, as well as other similar compounds, are very promising concepts for improving cell viability during the functionalization period required by nearly all tissue engineered constructs.

6   Conclusions

Microextrusion, to date, has been used to fabricate an amazing number of biologically relevant constructs. Ranging from surgical/anatomical models to organs-on-a-chip to cardiac patches and bone scaffolds, bioprinting has demonstrated that it can be broadly applied to conventional medicine. However current limitations with regard to cell viability, nutrient transport, and print speeds have confined clinical applications to drug testing (using organ-on-a-chip technology), small constructs, and surgical models. While each of these have proved beneficial and contributed to personalized health care, the fact remains that today's bioprinters cannot create full scale human tissues.

Before microextrusion printing can truly reach the pinnacle that is fully functional tissues, researchers will need to address current deficiencies in appropriate bioinks, vascularization, and prolonged cell viability. As the field progresses we expect to see continued development of bioinks. These will likely be highly varied to suit specific tissue types but we hypothesize that advances will be made by incorporating specific growth factors and further optimization of mechanical properties mirroring those found in the native tissue. Specifically work on gels with tunable cross-linking mechanisms will be vital to ensuring appropriate rheological properties while maintaining integrity postprinting. Vascularization is likely to be achieved through a combination of distinct techniques—specifically the combination of prepatterned vessels with self-assembled ones. Necessary for solid organs/tissues, the finite resolution imposed on microextrusion by nozzle size and the associated cell viability dictates that capillaries and microcapillaries will not be directly printed, but there is evidence that the development of these systems can be supported by larger microextruded vessels for a gradated vascular network. Finally, though a printed vasculature will provide support to a construct, progress must be made to ensure cell viability throughout the printing process. Again, this will likely be achieved through a combination of techniques. The duration of printing will need to be decreased—for complex components, this may be accomplished by modifying current extrusion printers to dispense multiple bioinks simultaneously or using multiple nozzles for the same bioink to dispense quickly over large areas. The incorporation of POGs will be vital in ensuring the central regions of large prints, and cells within the bioink reservoir are supplied with oxygen throughout the printing process until a media supply can be established.

The topics discussed within this chapter are, of course, not the only limiting factors keeping microextrusion printing from clinical relevance. Much work remains to be done with regard to obtaining and culturing cells from individual patients if graft rejection is to be circumvented. While current timelines are too slow and costly for mass use in the clinic, these are problems which plague all in vitro tissue fabrication systems equally. Though not part of the microextrusion process, it is important to acknowledge other steps of the fabrication process, such as sourcing and cell expansion, that will need to be optimized. This will likely also include alterations and new designs for specialized incubators and bioreactors that assist in maturing/supporting the tissue in vitro prior to implantation.

The ability to create suitably scaled tissues for translation into clinical use will revolutionize medicine. Bioprinting holds the key to the creation of individualized models, as has already been seen through the application of noncellular surgical models for preoperational visualization. Once researchers are able to print large cellularized constructs that maintain viability and functionality, they will be able to create healthy functional replacements for failed body parts. Initially, this is likely to encompass cartilage, skin and possibly solid organs including the kidney and liver. These systems alone would be a boon to society, drastically impacting the lives of patients waiting for transplants, those with arthritis, and victims of vehicular accidents and burns that have lost large portions of skin. However, as bioprinting continues to evolve and expand, including combining new techniques in innovative ways, there will be a day when researchers can replicate a required body part as necessary, drastically reducing morbidity and mortality within the population. While some may see this as science fiction, and it is true the technology has not yet reached the levels of a Star Trek replicator, one cannot deny the promise microextrusion printing holds in creating fully functional human tissues nor the revolutionary impact this development will have on conventional medicine. Current printers, despite their limitations and need for improvement, have already significantly impacted the field of bioengineering and have brought the ability to create replacement organs and tissues into view.

References

1. Khalil S, Nam J, Sun W (2005) Multi-nozzle deposition for construction of 3D biopolymer tissue scaffolds. Rapid Prototyp J 11(1):9–17. https://doi.org/10.1108/13552540510573347

2. Li M, Tian X, Zhu N, Schreyer DJ, Chen X (2010) Modeling process-induced cell damage in the biodispensing process. Tissue Eng Part C Methods 16(3):533–542. https://doi.org/10.1089/ten.TEC.2009.0178

3. Tirella A, Ahluwalia A (2012) The impact of fabrication parameters and substrate stiffness in direct writing of living constructs. Biotechnol Prog 28(5):1315–1320. https://doi.org/10.1002/btpr.1586

4. Buyukhatipoglu K, Jo W, Sun W, Alisa Morss C (2009) The role of printing parameters and scaffold biopolymer properties in the efficacy of a new hybrid nano-bioprinting system. Biofabrication 1(3):035003

5. Li M, Tian X, Schreyer DJ, Chen X (2011) Effect of needle geometry on flow rate and cell damage in the dispensing-based biofabrication process. Biotechnol Prog 27(6):1777–1784. https://doi.org/10.1002/btpr.679

6. Smith CM, Christian JJ, Warren WL, Williams SK (2007) Characterizing environmental factors that impact the viability of tissue-engineered constructs fabricated by a direct-write bioassembly tool. Tissue Eng 13(2):373–383. https://doi.org/10.1089/ten.2006.0101

7. Faulkner-Jones A, Greenhough S, King JA, Gardner J, Courtney A, Shu W (2013) Development of a valve-based cell printer for the formation of human embryonic stem cell spheroid aggregates. Biofabrication 5(1):015013

8. Reid JA, Mollica PA, Johnson GD, Ogle RC, Bruno RD, Sachs PC (2016) Accessible bioprinting: adaptation of a low-cost 3D-printer for precise cell placement and stem cell differentiation. Biofabrication 8(2):025017

9. Medicine WFIFR (2016) Printing skin cells on burn wounds. Wake Forest School of

Medicine.   http://www.wakehealth.edu/ Research/WFIRM/Research/Military-Applications/Printing-Skin-Cells-On-Burn-Wounds.htm. Accessed 14 Feb 2018

10. Hinton TJ, Jallerat Q, Palchesko RN, Park JH, Grodzicki MS, Shue H-J, Ramadan MH, Hudson AR, Feinberg AW (2015) Three-dimensional printing of complex biological structures by freeform reversible embedding of suspended hydrogels. Sci Adv 1(9): e1500758

11. Murphy SV, Atala A (2014) 3D bioprinting of tissues and organs. Nat Biotechnol 32:773. https://doi.org/10.1038/nbt.2958

12. Lee VK, Dias A, Ozturk MS, Chen K, Tricomi B, Corr DT, Intes X, Dai G (2015) 3D bioprinting and 3D imaging for stem cell engineering. In: Turksen K (ed) Bioprinting in regenerative Medicine. Springer International Publishing, Cham, pp 33–66. https://doi.org/10.1007/978-3-319-21386-6_2

13. Sears NA, Seshadri DR, Dhavalikar PS, Cosgriff-Hernandez E (2016) A review of three-dimensional printing in tissue engineering. Tissue Eng Part B Rev 22(4):298–310. https://doi.org/10.1089/ten.TEB.2015.0464

14. Billiet T, Gevaert E, De Schryver T, Cornelissen M, Dubruel P (2014) The 3D printing of gelatin methacrylamide cell-laden tissue-engineered constructs with high cell viability. Biomaterials 35(1):49–62. https://doi.org/10.1016/j.biomaterials.2013.09.078

15. Skardal A, Zhang J, McCoard L, Xu X, Oottamasathien S, Prestwich GD (2010) Photocrosslinkable Hyaluronan-gelatin hydrogels for two-step bioprinting. Tissue Eng A 16(8):2675–2685. https://doi.org/10.1089/ten.tea.2009.0798

16. Fielding GA, Bandyopadhyay A, Bose S (2012) Effects of silica and zinc oxide doping on mechanical and biological properties of 3D printed tricalcium phosphate tissue engineering scaffolds. Dent Mater 28(2):113–122. https://doi.org/10.1016/j.dental.2011.09.010

17. Malda J, Visser J, Melchels FP, Jüngst T, Hennink WE, Dhert WJA, Groll J, Hutmacher DW (2013) 25th anniversary article: engineering hydrogels for biofabrication. Adv Mater 25(36):5011–5028. https://doi.org/10.1002/adma.201302042

18. Dababneh AB, Ozbolat IT (2014) Bioprinting technology: a current state-of-the-art review. J Manuf Sci Eng 136(6):061016. https://doi.org/10.1115/1.4028512

19. Yerazunis WS (2016) Strengthening ABS, nylon, and polyester 3D printed parts by stress tensor aligned deposition paths and five-Axis printing. International Solid Freeform Fabrication Symposium

20. Grutle ØK (2015) 5-axis 3D printer. University of Oslo, Oslo

21. Kleefoot M (2016) ZHAW Master students develop novel 3D printers. Media Office ZHAW

22. Reinhardt D, Saunders R, Burry J (2016) Robotic fabrication in architecture, art and design 2016. Springer International Publishing, New York

23. Shi J (2014) Robotic extrusion (6-axis KUKA + ABS 3D printing). https://www.behance.net/gallery/22536831/ROBOTIC-EXTRUSION(6-Axis-KUKAABS-3D-Printing). Accessed 18 Feb 2018

24. NaGao E, Zhang G, Kizawa H (2015) 3D printed human liver-like tissue shows high level and sustained drug metabolic function. Third Annual Congress of Functional Analysis and Screening Technologies, Boston

25. Song X, Pan Y, Chen Y (2015) Development of a low-cost parallel kinematic machine for multidirectional additive manufacturing. J Manuf Sci Eng 137(2):021005. https://doi.org/10.1115/1.4028897

26. Thilmany J (2017) A Robot That Prints Tissue. The American Society of Mechanical Engineers. https://www.asme.org/engineering-topics/articles/bioengineering/a-robot-that-prints-tissue. Accessed 10 Oct 2017

27. Möller T, Amoroso M, Hägg D, Brantsing C, Rotter N, Apelgren P, Lindahl A, Kölby L, Gatenholm P (2017) In vivo Chondrogenesis in 3D bioprinted human cell-laden hydrogel constructs. Plast Reconstruct Surg Global Open 5(2):e1227. https://doi.org/10.1097/gox.0000000000001227

28. Geven MA, Sprecher C, Guillaume O, Eglin D, Grijpma DW (2017) Micro-porous composite scaffolds of photo-crosslinked poly(trimethylene carbonate) and nano-hydroxyapatite prepared by low-temperature extrusion-based additive manufacturing. Polym Adv Technol 28(10):1226–1232. https://doi.org/10.1002/pat.3890

29. Nguyen D, Hägg DA, Forsman A, Ekholm J, Nimkingratana P, Brantsing C, Kalogeropoulos T, Zaunz S, Concaro S, Brittberg M, Lindahl A, Gatenholm P, Enejder A, Simonsson S (2017) Cartilage tissue engineering by the 3D bioprinting of iPS cells in a Nanocellulose/alginate bioink. Sci

Rep 7(1):658. https://doi.org/10.1038/s41598-017-00690-y

30. En G, Barruet E, Hsiao E, Schepers K, Presnell S, Nguyen D, Retting K (2014) Three-dimensional (3D) bone tissues derived from stem cells as a novel model for mineralization. Stem Cell Meeting on the Mesa, La Jolla, CA

31. Hardwick RN, Nguyen DG, Robbins J, Grundy C, Gorgen V, Bangalore P, Perusse D, Creasey O, King S, Lin S, Khatiwala C, Halberstadt C, Presnell SC (2015) Functional characterization of three-dimensional (3D) human liver tissues generated by an automated bioprinting platform. Boston, MA, American Society of Experimental Biology

32. Nguyen DG, Funk J, Robbins JB, Crogan-Grundy C, Presnell SC, Singer T, Roth AB (2016) Bioprinted 3D primary liver tissues allow assessment of organ-level response to clinical drug induced toxicity in vitro. PLoS One 11(7):e0158674. https://doi.org/10.1371/journal.pone.0158674

33. Machino R, Matsumoto K, Taura Y, Yamasaki N, Tagagi K, Tsuchiya T, Miyazaki T, Nakayama K, Nagayasu T (2015) Scaffold-free trachea tissue engineering using bioprinting. Am J Respir Crit Care Med 191:A5343

34. Visscher DO, Bos EJ, Peeters M, Kuzmin NV, Groot ML, Helder MN, van Zuijlen PPM (2016) Cartilage tissue engineering: preventing tissue scaffold contraction using a 3D-printed polymeric cage. Tissue Eng Part C Methods 22(6):573. https://doi.org/10.1089/ten.TEC.2016.0073

35. King SM, Gorgen V, Presnell SC, Nguyen DG, Sheperd BR (2013) Development of 3D bioprinted human breast cancer for in vitro screening of therapeutics targeted against cancer progression. American Society of Biology, New Orleans, LA

36. Zhang YS, Davoudi F, Walch P, Manbachi A, Luo X, Dell'Erba V, Miri AK, Albadawi H, Arneri A, Li X, Wang X, Dokmeci MR, Khademhosseini A, Oklu R (2016) Bioprinted thrombosis-on-a-chip. Lab Chip 16 (21):4097–4105. https://doi.org/10.1039/C6LC00380J

37. Hanumegowda UM, Wu Y, Smith TR, Lehnam-McKeeman L (2016) Monocrotaline toxicity in 3D-bioprinted human liver tissue. Society of Toxicology Meeting, New Orleans, LA

38. Hardwick RN, Hampton J, Perusse D, Nguyen D (2016) Inflammatory response of Kupffer cells in 3D bioprinted human liver tissues. Society of Toxicology Meeting, New Orleans, LA

39. Norona LM, Nguyen DG, Gerber DA, Presnell SC, LeCluyse EL (2016) Editor's highlight: modeling compound-induced fibrogenesis in vitro using three-dimensional bioprinted human liver tissues. Toxicol Sci 154(2):354–367. https://doi.org/10.1093/toxsci/kfw169

40. Sultan S, Siqueira G, Zimmermann T, Mathew AP (2017) 3D printing of nano-cellulosic biomaterials for medical applications. Curr Opin Biomed Eng 2(Supplement C):29–34. https://doi.org/10.1016/j.cobme.2017.06.002

41. Mendoza HR (2015) 3D Bioprinting Takes Another Step Forward with the Launch of First Living Cellular Bioink Kits. https://3dprint.com/97382/first-living-cellular-bioink/. Accessed 10 Oct 2017

42. Ersumo N, Witherel CE, Spiller KL (2016) Differences in time-dependent mechanical properties between extruded and molded hydrogels. Biofabrication 8(3):035012

43. Dubbin K, Hori Y, Lewis KK, Heilshorn SC (2016) Dual-stage crosslinking of a gel-phase bioink improves cell viability and homogeneity for 3D bioprinting. Adv Healthc Mater 5 (19):2488–2492. https://doi.org/10.1002/adhm.201600636

44. Xconomy: TeVido Taps 3D Printing Technology to Build a Better Nipple (2014). Newstex, Chatham

45. Horváth L, Umehara Y, Jud C, Blank F, Petri-Fink A, Rothen-Rutishauser B (2015) Engineering an in vitro air-blood barrier by 3D bioprinting. Sci Rep 5:7974. https://doi.org/10.1038/srep07974

46. Itoh M, Nakayama K, Noguchi R, Kamohara K, Furukawa K, Uchihashi K, Toda S, Oyama J-I, Node K, Morita S (2015) Scaffold-free tubular tissues created by a bio-3D printer undergo remodeling and endothelialization when implanted in rat aortae. PLoS One 10(9):e0136681. https://doi.org/10.1371/journal.pone.0136681

47. Yamamoto T, Funahashi Y, Mastukawa Y, Tsuji Y, Mizuno H, Nakayama K, Gotoh M (2015) MP19–17 Human urethra-engineered with human mesenchymal stem cell with maturation by rearrangement of cells for self-organization—newly developed scaffold-free three-dimensional bio-printer. J Urol 193(4):e221–e222. https://doi.org/10.1016/j.juro.2015.02.1009

48. Chang R, Nam J, Sun W (2008) Effects of dispensing pressure and nozzle diameter on

cell survival from solid freeform fabrication-based direct cell writing. Tissue Eng A 14 (1):41–48. https://doi.org/10.1089/ten.a.2007.0004

49. Faulkner-Jones A, Fyfe C, Cornelissen DJ, Gardner J, King J, Courtney A, Shu W (2015) Bioprinting of human pluripotent stem cells and their directed differentiation into hepatocyte-like cells for the generation of mini-livers in 3D. Biofabrication 7 (4):044102. https://doi.org/10.1088/1758-5090/7/4/044102

50. Kim G, Ahn S, Yoon H, Kim Y, Chun W (2009) A cryogenic direct-plotting system for fabrication of 3D collagen scaffolds for tissue engineering. J Mater Chem 19 (46):8817–8823. https://doi.org/10.1039/B914187A

51. Lee YB, Polio S, Lee W, Dai G, Menon L, Carroll RS, Yoo SS (2010) Bio-printing of collagen and VEGF-releasing fibrin gel scaffolds for neural stem cell culture. Exp Neurol 223(2):645–652. https://doi.org/10.1016/j.expneurol.2010.02.014

52. Markstedt K, Sundberg J, Gatenholm P (2014) 3D bioprinting of cellulose structures from an ionic liquid. 3D Print Addit Manufact 1(3):115–121. https://doi.org/10.1089/3dp.2014.0004

53. Wenz A, Janke K, Hoch E, Tovar Günter EM, Borchers K, Kluger Petra J (2016) Hydroxyapatite-modified gelatin bioinks for bone bioprinting. BioNanoMaterials 17. https://doi.org/10.1515/bnm-2015-0018

54. Zhang Y, Yu Y, Chen H, Ozbolat IT (2013) Characterization of printable cellular microfluidic channels for tissue engineering. Biofabrication 5(2):025004. https://doi.org/10.1088/1758-5082/5/2/025004

55. Park JY, Choi JC, Shim JH, Lee JS, Park H, Kim SW, Doh J, Cho DW (2014) A comparative study on collagen type I and hyaluronic acid dependent cell behavior for osteochondral tissue bioprinting. Biofabrication 6 (3):035004. https://doi.org/10.1088/1758-5082/6/3/035004

56. Markstedt K, Mantas A, Tournier I, Martinez Avila H, Hagg D, Gatenholm P (2015) 3D bioprinting human chondrocytes with Nanocellulose-alginate bioink for cartilage tissue engineering applications. Biomacromolecules 16(5):1489–1496. https://doi.org/10.1021/acs.biomac.5b00188

57. Vanderhooft JL, Alcoutlabi M, Magda JJ, Prestwich GD (2009) Rheological properties of cross-linked hyaluronan-gelatin hydrogels for tissue engineering. Macromol Biosci 9 (1):20–28. https://doi.org/10.1002/mabi.200800141

58. Paxton NC, Smolan W, Bock T, Melchels FPW, Groll J, Juengst T (2017) Proposal to assess printability of bioinks for extrusion-based bioprinting and evaluation of rheological properties governing bioprintability. Biofabrication 9:044107. https://doi.org/10.1088/1758-5090/aa8dd8

59. Schuurman W, Levett PA, Pot MW, van Weeren PR, Dhert WJ, Hutmacher DW, Melchels FP, Klein TJ, Malda J (2013) Gelatin-methacrylamide hydrogels as potential biomaterials for fabrication of tissue-engineered cartilage constructs. Macromol Biosci 13 (5):551–561. https://doi.org/10.1002/mabi.201200471

60. Burdick JA, Chung C, Jia X, Randolph MA, Langer R (2005) Controlled degradation and mechanical behavior of Photopolymerized hyaluronic acid networks. Biomacromolecules 6(1):386–391. https://doi.org/10.1021/bm049508a

61. Poldervaart MT, Goversen B, de Ruijter M, Abbadessa A, Melchels FPW, Öner FC, Dhert WJA, Vermonden T, Alblas J (2017) 3D bioprinting of methacrylated hyaluronic acid (MeHA) hydrogel with intrinsic osteogenicity. PLoS One 12(6):e0177628. https://doi.org/10.1371/journal.pone.0177628

62. Loessner D, Meinert C, Kaemmerer E, Martine LC, Yue K, Levett PA, Klein TJ, Melchels FP, Khademhosseini A, Hutmacher DW (2016) Functionalization, preparation and use of cell-laden gelatin methacryloyl-based hydrogels as modular tissue culture platforms. Nat Protoc 11(4):727–746. https://doi.org/10.1038/nprot.2016.037

63. Van Den Bulcke AI, Bogdanov B, De Rooze N, Schacht EH, Cornelissen M, Berghmans H (2000) Structural and rheological properties of methacrylamide modified gelatin hydrogels. Biomacromolecules 1 (1):31–38

64. Sutter M, Siepmann J, Hennink WE, Jiskoot W (2007) Recombinant gelatin hydrogels for the sustained release of proteins. J Control Release 119(3):301–312. https://doi.org/10.1016/j.jconrel.2007.03.003

65. Benton JA, DeForest CA, Vivekanandan V, Anseth KS (2009) Photocrosslinking of gelatin macromers to synthesize porous hydrogels that promote valvular interstitial cell function. Tissue Eng A 15(11):3221–3230. https://doi.org/10.1089/ten.TEA.2008.0545

66. Aubin H, Nichol JW, Hutson CB, Bae H, Sieminski AL, Cropek DM, Akhyari P, Khademhosseini A (2010) Directed 3D cell

alignment and elongation in microengineered hydrogels. Biomaterials 31(27):6941–6951. https://doi.org/10.1016/j.biomaterials. 2010.05.056

67. Nichol JW, Koshy ST, Bae H, Hwang CM, Yamanlar S, Khademhosseini A (2010) Cell-laden microengineered gelatin methacrylate hydrogels. Biomaterials 31(21):5536–5544. https://doi.org/10.1016/j.biomaterials. 2010.03.064

68. Qi H, Du Y, Wang L, Kaji H, Bae H, Khademhosseini A (2010) Patterned differentiation of individual Embryoid bodies in spatially organized 3D hybrid microgels. Adv Mater 22(46):5276–5281. https://doi.org/10.1002/adma.201002873

69. Chen YC, Lin RZ, Qi H, Yang Y, Bae H, Melero-Martin JM, Khademhosseini A (2012) Functional human vascular network generated in Photocrosslinkable gelatin methacrylate hydrogels. Adv Funct Mater 22 (10):2027–2039. https://doi.org/10.1002/adfm.201101662

70. Camci-Unal G, Cuttica D, Annabi N, Demarchi D, Khademhosseini A (2013) Synthesis and characterization of hybrid hyaluronic acid-gelatin hydrogels. Biomacromolecules 14(4):1085–1092. https://doi.org/10.1021/bm3019856

71. Skardal A, Devarasetty M, Kang HW, Mead I, Bishop C, Shupe T, Lee SJ, Jackson J, Yoo J, Soker S, Atala A (2015) A hydrogel bioink toolkit for mimicking native tissue biochemical and mechanical properties in bioprinted tissue constructs. Acta Biomater 25:24–34. https://doi.org/10.1016/j.actbio.2015.07. 030

72. Merceron TK, Burt M, Seol YJ, Kang HW, Lee SJ, Yoo JJ, Atala A (2015) A 3D bioprinted complex structure for engineering the muscle-tendon unit. Biofabrication 7 (3):035003. https://doi.org/10.1088/1758-5090/7/3/035003

73. Kaully T, Kaufman-Francis K, Lesman A, Levenberg S (2009) Vascularization-the conduit to viable engineered tissues. Tissue Eng Part B: Rev 15(2):159–169. https://doi.org/10.1089/ten.teb.2008.0193

74. Bertassoni LE, Cecconi M, Manoharan V, Nikkhah M, Hjortnaes J, Cristino AL, Barabaschi G, Demarchi D, Dokmeci MR, Yang YZ, Khademhosseini A (2014) Hydrogel bioprinted microchannel networks for vascularization of tissue engineering constructs. Lab Chip 14(13):2202–2211. https://doi.org/10.1039/c4lc00030g

75. Nashimoto Y, Hayashi T, Kunita I, Nakamasu A, Torisawa Y, Nakayama M, Takigawa-Imamura H, Kotera H, Nishiyama K, Miura T, Yokokawa R (2017) Integrating perfusable vascular networks with a three-dimensional tissue in a microfluidic device. Integr Biol 9(6):506–518. https://doi.org/10.1039/c7ib00024c

76. Moya M, Tran D, George SC (2013) An integrated in vitro model of perfused tumor and cardiac tissue. Stem Cell Res Ther 4: S15–S15. https://doi.org/10.1186/scrt376

77. Gao G, Lee JH, Jang J, Lee DH, Kong JS, Kim BS, Choi YJ, Jang WB, Hong YJ, Kwon SM, Cho DW (2017) Tissue engineered bio-blood-vessels constructed using a tissue-specific bioink and 3D coaxial cell printing technique: a novel therapy for ischemic disease. Adv Funct Mater 27(33):1700798. https://doi.org/10.1002/adfm.201700798

78. Ju YM, Choi JS, Atala A, Yoo JJ, Lee SJ (2010) Bilayered scaffold for engineering cellularized blood vessels. Biomaterials 31 (15):4313–4321. https://doi.org/10.1016/j.biomaterials.2010.02.002

79. Assmann A, Delfs C, Munakata H, Schiffer F, Horstkötter K, Huynh K, Barth M, Stoldt VR, Kamiya H, Boeken U, Lichtenberg A, Akhyari P (2013) Acceleration of autologous in vivo recellularization of decellularized aortic conduits by fibronectin surface coating. Biomaterials 34(25):6015–6026. https://doi.org/10.1016/j.biomaterials.2013.04. 037

80. Dall'Olmo L, Zanusso I, Di Liddo R, Chioato T, Bertalot T, Guidi E, Conconi MT (2014) Blood vessel-derived acellular matrix for vascular graft application. Biomed Res Int 2014:685426. https://doi.org/10. 1155/2014/685426

81. Jung Y, Ji H, Chen Z, Fai Chan H, Atchison L, Klitzman B, Truskey G, Leong KW (2015) Scaffold-free, human mesenchymal stem cell-based tissue engineered blood vessels. Sci Rep 5:15116. https://doi.org/10. 1038/srep15116. https://www.nature.com/articles/srep15116#supplementary-information

82. L'Heureux N, Paquet S, Labbe R, Germain L, Auger FA (1998) A completely biological tissue-engineered human blood vessel. FASEB J 12(1):47–56

83. Bos GW, Poot AA, Beugeling T, van Aken WG, Feijen J (1998) Small-diameter vascular graft prostheses: current status. Arch Physiol Biochem 106(2):15

84. Seifu DG, Purnama A, Mequanint K, Mantovani D (2013) Small-diameter vascular tissue engineering. Nat Rev Cardiol 10:410. https://doi.org/10.1038/nrcardio.2013.77

85. McAllister TN, Maruszewski M, Garrido SA, Wystrychowski W, Dusserre N, Marini A, Zagalski K, Fiorillo A, Avila H, Manglano X, Antonelli J, Kocher A, Zembala M, Cierpka L, de la Fuente LM, L'Heureux N (2009) Effectiveness of haemodialysis access with an autologous tissue-engineered vascular graft: a multicentre cohort study. Lancet 373 (9673):1440–1446. https://doi.org/10.1016/s0140-6736(09)60248-8

86. L'Heureux N, McAllister TN, de la Fuente LM (2007) Tissue-engineered blood vessel for adult arterial revascularization. N Engl J Med 357(14):1451–1453. https://doi.org/10.1056/NEJMc071536

87. Suntornnond R, An J, Kai Chua C (2017) Roles of support materials in 3D bioprinting—present and future, vol 3. doi:https://doi.org/10.18063/IJB.2017.01.006

88. Miller JS, Stevens KR, Yang MT, Baker BM, Nguyen D-HT, Cohen DM, Toro E, Chen AA, Galie PA, Yu X, Chaturvedi R, Bhatia SN, Chen CS (2012) Rapid casting of patterned vascular networks for perfusable engineered three-dimensional tissues. Nat Mater 11:768. https://doi.org/10.1038/nmat3357. https://www.nature.com/articles/nmat3357#supplementary-information

89. Liu J, Li Y, Fan H, Zhu Z, Jiang J, Ding R, Hu Y, Huang X (2010) Iron oxide-based nanotube arrays derived from sacrificial template-accelerated hydrolysis: large-area design and reversible Lithium storage. Chem Mater 22(1):212–217. https://doi.org/10.1021/cm903099w

90. Kolesky DB, Homan KA, Skylar-Scott MA, Lewis JA (2016) Three-dimensional bioprinting of thick vascularized tissues. Proc Natl Acad Sci 113(12):3179

91. Nie S, Hsiao WLW, Pan W, Yang Z (2011) Thermoreversible Pluronic(®) F127-based hydrogel containing liposomes for the controlled delivery of paclitaxel: in vitro drug release, cell cytotoxicity, and uptake studies. Int J Nanomedicine 6:151–166. https://doi.org/10.2147/IJN.S15057

92. Kolesky DB, Truby RL, Gladman AS, Busbee TA, Homan KA, Lewis JA (2014) 3D bioprinting of vascularized, heterogeneous cell-laden tissue constructs. Adv Mater 26 (19):3124–3130. https://doi.org/10.1002/adma.201305506

93. Homan KA, Kolesky DB, Skylar-Scott MA, Herrmann J, Obuobi H, Moisan A, Lewis JA (2016) Bioprinting of 3D convoluted renal proximal tubules on Perfusable chips. Sci Rep 6:34845. https://doi.org/10.1038/srep34845. https://www.nature.com/articles/srep34845#supplementary-information

94. Wu W, DeConinck A, Lewis JA (2011) Omnidirectional printing of 3D microvascular networks. Adv Mater 23(24):H178–H183. https://doi.org/10.1002/adma.201004625

95. Wilson WC, Boland T (2003) Cell and organ printing 1: protein and cell printers. The anatomical record part a: discoveries in molecular, cellular. Evol Biol 272A(2):491–496. https://doi.org/10.1002/ar.a.10057

96. Boland T, Mironov V, Gutowska A, Roth EA, Markwald RR (2003) Cell and organ printing 2: fusion of cell aggregates in three-dimensional gels. Anat Rec A Discov Mol Cell Evol Biol 272(2):497–502

97. Christensen K, Xu C, Chai W, Zhang Z, Fu J, Huang Y (2015) Freeform inkjet printing of cellular structures with bifurcations. Biotechnol Bioeng 112(5):1047–1055. https://doi.org/10.1002/bit.25501

98. Jia W, Gungor-Ozkerim PS, Zhang YS, Yue K, Zhu K, Liu W, Pi Q, Byambaa B, Dokmeci MR, Shin SR, Khademhosseini A (2016) Direct 3D bioprinting of perfusable vascular constructs using a blend bioink. Biomaterials 106:58–68. https://doi.org/10.1016/j.biomaterials.2016.07.038

99. Gao Q, He Y, Fu JZ, Liu A, Ma L (2015) Coaxial nozzle-assisted 3D bioprinting with built-in microchannels for nutrients delivery. Biomaterials 61:203–215. https://doi.org/10.1016/j.biomaterials.2015.05.031

100. Zhang Y, Yu Y, Ozbolat IT (2013) Direct bioprinting of vessel-like tubular microfluidic channels. J Nanotechnol Eng Med 4 (2):0210011–0210017. https://doi.org/10.1115/1.4024398

101. Zhang Y, Yu Y, Akkouch A, Dababneh A, Dolati F, Ozbolat IT (2015) In vitro study of directly bioprinted Perfusable vasculature conduits. Biomater Sci 3(1):134–143. https://doi.org/10.1039/c4bm00234b

102. Yu Y, Zhang Y, Martin JA, Ozbolat IT (2013) Evaluation of cell viability and functionality in vessel-like bioprintable cell-laden tubular channels. J Biomech Eng 135(9):91011

103. Attalla R, Ling C, Selvaganapathy P (2016) Fabrication and characterization of gels with integrated channels using 3D printing with microfluidic nozzle for tissue engineering applications. Biomed Microdevices 18(1):17. https://doi.org/10.1007/s10544-016-0042-6

104. Riemenschneider SB, Mattia DJ, Wendel JS, Schaefer JA, Ye L, Guzman PA, Tranquillo RT (2016) Inosculation and perfusion of

pre-vascularized tissue patches containing aligned human microvessels after myocardial infarction. Biomaterials 97:51–61. https://doi.org/10.1016/j.biomaterials.2016.04.031

105. Sobrino A, Phan DTT, Datta R, Wang XL, Hachey SJ, Romero-Lopez M, Gratton E, Lee AP, George SC, Hughes CCW (2016) 3D microtumors in vitro supported by perfused vascular networks. Sci Rep 6:31589. https://doi.org/10.1038/srep31589

106. Fedorovich NE, De Wijn JR, Verbout AJ, Alblas J, Dhert WJ (2008) Three-dimensional fiber deposition of cell-laden, viable, patterned constructs for bone tissue printing. Tissue Eng Part A 14(1):127–133

107. Hendriks J, Willem Visser C, Henke S, Leijten J, Saris DBF, Sun C, Lohse D, Karperien M (2015) Optimizing cell viability in droplet-based cell deposition. Sci Rep 5:11304. https://doi.org/10.1038/srep11304

108. Kim J, Andersson KE, Jackson JD, Lee SJ, Atala A, Yoo JJ (2014) Downregulation of metabolic activity increases cell survival under hypoxic conditions: potential applications for tissue engineering. Tissue Eng A 20 (15–16):2265–2272. https://doi.org/10.1089/ten.TEA.2013.0637

109. Buck LT (2004) Adenosine as a signal for ion channel arrest in anoxia-tolerant organisms. Comp Biochem Physiol B Biochem Mol Biol 139(3):401–414. https://doi.org/10.1016/j.cbpc.2004.04.002

Chapter 6

Stereolithography 3D Bioprinting

Hitendra Kumar and Keekyoung Kim

Abstract

Stereolithography (SLA) 3D bioprinting has emerged as a prominent bioprinting method addressing the requirements of complex tissue fabrication. This chapter addresses the advancement in SLA 3D bioprinting in concurrent with the development of novel photocrosslinkable biomaterials with enhanced physical and chemical properties. We discuss the cytocompatible photoinitiators operating in the wide spectrum of the ultraviolet (UV) and the visible light and high-resolution dynamic mask projection systems with a suitable illumination source. The potential of SLA 3D bioprinting has been explored in various themes, like bone and neural tissue engineering and in the development of controlled microenvironments to study cell behavior. The flexible design and versatility of SLA bioprinting makes it an attractive bioprinting process with myriad possibilities and clinical applications.

Key words Stereolithography 3D bioprinting, Photocrosslinking, Hydrogel scaffolds, Tissue and organ regeneration

1 Introduction

Stereolithography (SLA) 3D bioprinting has emerged as a prominent bioprinting method addressing the requirements of complex scaffold tissue fabrication. Compared to conventional bioprinting methods, the SLA 3D bioprinting offers several advantages in the form of multiscalability, high resolution, and rapid printing of highly complex scaffolds. In addition, the ability to fabricate implantable scaffolds with anatomically accurate geometry, precisely controlled surface characteristics, and tunable physical and chemical properties make it highly relevant for clinical applications.

Similar to the SLA fabrication method used in microfabrication or 3D printing, the SLA bioprinting fabricates 3D tissue constructs in a layer-by-layer manner. For this purpose, the designed 3D model is converted into a stack of the sliced 2D images. Each 2D image in the stack represents the cross-section of the 3D solid model at a specified plane from the top. Hence, the input for the SLA 3D bioprinting process is in the form of slice images. Many biomedical imaging methods, such as computed tomography

Jeremy M. Crook (ed.), *3D Bioprinting: Principles and Protocols*, Methods in Molecular Biology, vol. 2140,
https://doi.org/10.1007/978-1-0716-0520-2_6, © Springer Science+Business Media, LLC, part of Springer Nature 2020

(CT) and magnetic resonance imaging (MRI), generate tissue images in the form of image stacks. Based on this similarity, biomedical imaging data can be directly utilized in the SLA bioprinting. Hence, the SLA 3D bioprinting can be used for a number of applications based on the wide range of input datasets—from primitive user-defined geometries to complex tissue structures. Some of these applications are discussed in the later sections of this chapter.

Other important components of the SLA 3D bioprinting system are illumination sources and bioinks. The illumination source is the key element to define the performance of the SLA process. The resolution and accuracy of the printed parts are often determined by the characteristics of the illumination source. Moreover, the mechanical properties of the scaffold can also be tuned by controlling the illumination conditions [1]. Further discussion about the characteristics of the illumination sources is presented in subsequent sections.

In addition to the illumination source, bioinks are raw materials for the bioprinting process and consist of a biomaterial mixed with cells. The choice of the biomaterials, as well as the cell types, is guided by the intended application. The biomaterials are required to possess certain properties specific to the bioprinting process. In the case of the extrusion-based bioprinting, the biomaterials should be capable of undergoing thermal gelation at room temperature or should be chemically cross-linkable before or during the material extrusion [2]. For the SLA bioprinting, the biomaterials should be photocrosslinkable and, therefore, the bioink solution should be cross-linked by the illumination of light.

In an overview, the SLA 3D bioprinting process begins with the preparation of a suitable bioink. The bioink is prepared by mixing cells in a photosensitive prepolymer solution. Next, the 3D model to be printed is converted into a stack of slice images and provided as the input for the bioprinting system. In the fabrication steps, first, the bioink is dispensed in a petri dish to achieve a desired layer thickness of the material. Second, an image from the input image stack is used as the mask and projected onto the dispensed bioink to begin cross-linking. These two steps of dispensing the bioink and projecting the patterned image are repeated until all the patterned images have been projected to build a 3D construct. Finally, the uncrosslinked portion of the bioink is removed to obtain the 3D printed scaffold with encapsulated cells.

2   Materials for SLA 3D Bioprinting

As mentioned in the previous section, the bioink consists of the biomaterial prepolymer solution and the cells. Although, there have been some studies to directly use an assembly of cells to build 3D constructs, the most common way to build 3D tissues is to use hydrogels as scaffolding materials [3–6]. There are several advantages associated with the use of hydrogels. For instance, the hydrogels provide an excellent supporting structure for cells during culturing in 3D. For the purpose of SLA bioprinting, the hydrogel should be formed by a pre-polymer solution which is cross-linkable upon interaction with light. Wide spectrums are used in the source of the illumination for cross-linking. Commonly used systems consist of UV or visible light as required by the selected photoinitiator. To undergo photocrosslinking, two commonly used mechanisms are acryloyl-based photocrosslinking and thiol–ene click reaction. Hence, hydrogel macromers consist of either acryloyl functional group or an alkenyl functional group. The cytotoxicity and reactivity of the functional group should be considered to determine the type and number of reactive groups introduced in the macromer backbone to ensure good biocompatibility of the resulting hydrogels [7].

2.1 Photocross-linkable Hydrogels

Acryloyl or acrylate hydrogels are a class of most commonly used photocrosslinkable biomaterials in tissue engineering and biomedical applications. These hydrogels contain an electron-deficient functional group with terminal carbon–carbon double bond. Methacryloyl-based photocrosslinkable hydrogels are synthesized from macromolecular hydrogels by reacting with a methacrylation agent (e.g., glycidyl methacrylate or methacrylic anhydride) to substitute an acryloyl group (Fig. 1). The substitution degree of methacryloyl functional groups can be controlled by the concentration of the methacrylation agent, the duration of reaction, and the pH of the medium during the synthesis [8, 9]. Some of the hydrogels falling into this category are poly (ethylene glycol) diacrylate (PEGDA), poly (ethylene glycol) dimethacrylate (PEGDMA), gelatin methacryloyl (GelMA), and dextran methacrylate (DexMA). The acryloyl hydrogels undergo chain polymerization reaction in the presence of free radicals to form a cross-linked network of hydrogels [10]. Due to excellent biocompatibility, gelatin-based hydrogels have been widely used in SLA 3D bioprinting [10, 11]. By using different photoinitiators, complex GelMA scaffolds were fabricated by UV, visible light, and two-photon based SLA bioprinting systems [9, 10, 12–19]. Soman et al. used a maskless SLA printing system to fabricate complex microscale free-form structures of flower, pyramid, spiral, and dome shapes [20]. Gauvin et al. also used a projection SLA bioprinting system to fabricate 3D cell-encapsulated scaffolds using GelMA and demonstrated high cell viability and cell elongation. Further, the well-controlled porosity of the scaffold showed the development of vascular networks [21].

From the perspective of the SLA bioprinting, norbornene hydrogels have demonstrated high potential. They provide an alternative to acryloyl hydrogels and are not affected by the problem of oxygen inhibition. Norbornene hydrogels also form cross-linked

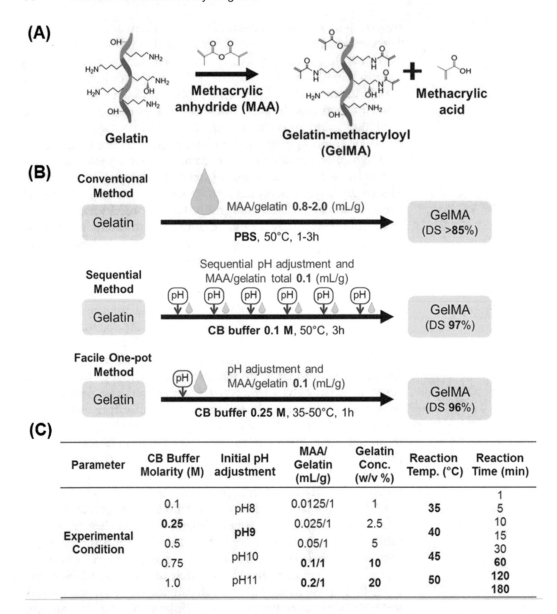

Fig. 1 (a) Schematic representation of gelatin-methacryloyl (GelMA) synthesis. (b) Representation of different synthesis conditions with varying amounts of methacrylate anhydride (MAAnh) and resulting degree of substitution. (c) Synthesis parameters and optimized conditions. (Adapted from Shirahama et al., *Scientific reports 6*, 2016)

networks of uniform cross-linking density and fewer network defects which allows for better tuning and predictability of hydrogel properties [7, 22]. Norbornene hydrogels have also gained importance in recent years due to photocrosslinking in presence of visible light [23]. As the name suggests, this photocrosslinking mechanism involves the participation of a thiol group and reactive carbon–carbon double bonds, termed as "enes." The reaction propagates by either a free radical, which is called as a thiol–ene reaction or an anionic chain, which is called as thiol Michael addition reaction [24]. Types of alkenes used for thiol–ene hydrogel formation are vinyl sulfone, maleimides, norbornene, and acryloyl groups. Vinyl sulfone groups are introduced by the reaction of macromers with hydroxyl functional groups and free amine groups. These are electron deficient alkenes that react with thiols to produce stable β-thiosulfonyl linkages, also known as thioether bonds [25, 26]. In recent research works, vinyl sulfone functional groups have been introduced to PEG hydrogels and dextran to initiate thiol–ene click reaction [25]. These hydrogels were utilized for various biomedical applications and drug delivery [27–30]. Norbornene is a strained cyclohexane ring with methylene bridge and reacts with thiols in presence of free radicals [25] Norbornene has been used with photoinitiators 2-hydroxy-4′-(2-hydroxyethoxy)-2-methylpropiophenone (Irgacure 2959) for photocrosslinking by a low dosage of UV-A light. Gelatin with norbornene functional groups (GelNB) has been used with visible light photoinitiator 2′,4′,5′,7′-tetrabromofluorescein (Eosin-Y) and lithium phenyl-2,4,6-trimethylbenzoylphosphinate (LAP) for cross-linking by visible light (400–700 nm) [31, 32].

2.2 Photoinitiators

Photocrosslinking reaction is the main component of SLA 3D bioprinting process. In previous subsections, we discussed various types of photocrosslinkable hydrogels used for tissue engineering. Photoinitiators cleave to form a free radical and initiate the photocrosslinking reaction of hydrogels upon excitation by light with suitable wavelengths. Photoinitiators are typically classified into Type-I and Type-II based on intermediate steps involved in the cross-linking.

Type-I photoinitiators are often referred as cleavable photoinitiators and do not require multiple components. When the light of suitable wavelength span is illuminated, Type-I photoinitiators form an excited triplet state. This state is short-lived and then undergoes cleavage to produce two free radicals which initiate the cross-linking of hydrogel [33]. Irgacure 2959 is the most widely used Type-I photoinitiator with good cytocompatibility [9, 21, 34]. LAP is another Type-I photoinitiator which can be used with both UV-A (365 nm) and visible light (405 nm) wavelengths. With high water solubility, high polymerization rate, and better cytocompatibility, LAP became more preferable than commonly used Irgacure 2959 [35].

On the other hand, Type-II photoinitiators are referred to as bimolecular photoinitiating systems (e.g., benzophenone/tertiary amine). The free radical formation by Type-II systems begins with the excitation of benzophenone which initiates a fast electron transfer from the tertiary amine. This step is followed by a slow proton transfer process resulting in the formation of H-donor radical for initiating the cross-linking of hydrogels [33]. Involvement of additional steps increase the complexity of the cross-linking mechanism and, on the other hand, decrease the efficiency of Type-II systems which makes them slower than Type-I systems [36]. Several studies have utilized eosin-Y-based visible light photocrosslinking and demonstrated high cytocompatibility for 3D bioprinting applications [11, 18, 23, 32, 37, 38].

The SLA process is a relatively rapid 3D bioprinting process, but the resolution and time required for fabricating 3D tissue constructs depends on the number of projection layers. Hence, the bioprinting process often may extend for more than an hour of light exposure, resulting in lower cell survival rate using UV light-based photoinitiation than visible light-based photoinitiation [34].

3   Classification of SLA 3D Bioprinting

The photocrosslinking process has been utilized to develop different types of SLA 3D bioprinting systems. The SLA 3D bioprinting systems can be classified into several types based on the illumination source and pattern projection method. Single-photon and multiphoton methods are two types of SLA 3D bioprinting systems based on illumination source. In the single-photon method, one photon of high energy is absorbed by the photoinitiator to generate free radicals. Visible light and UV-based SLA 3D bioprinting systems fall under this category. In the multiphoton or two-photon method, multiple low energy photons are sequentially absorbed by the photoinitiator in the excited transition state to generate free radicals [39]. Infrared (IR) illumination is a common illumination source used for two-photon bioprinting systems [19, 40]. Digital light processing (DLP) projection is another category of SLA 3D bioprinting systems based on the pattern projection method.

3.1 DLP Projection SLA 3D Bioprinting

The DLP system, also known as a dynamic projection or maskless projection system, consists of a digital micromirror device (DMD) for the projection of the patterned light that is also referred to as a dynamic mask [41–43]. A DMD is an array of microscale mirrors, which are typically in the size of 5–10 μm and undergo rotation and manipulate the direction of reflected light [44]. In a DLP system, a lamp is used as a high-intensity illumination source with a wide spectrum. The light from lamp is incident on the DMD array. A graphical processing unit (GPU) reads the input digital mask and, through a controller, applies a potential to the DMD array to manipulate the orientation of individual mirrors. Light is reflected by the DMD array, and the orientation of the mirrors determines the intensity of reflected light along the projection direction. A major advantage of DLP systems is the ability to project any desired pattern. The microscale mirrors in DMD array provide a very high-resolution pattern projection which can be further enhanced by using focusing lenses. Nearly all SLA 3D bioprinting systems utilize DMD based dynamic mask for fabrication of macro and microscale scaffolds [18, 19, 21, 42, 45–54].

The DLP-based SLA 3D bioprinting has gained spotlight since the past decade due to its ability to print complex structures. UV illumination has been used in the majority of SLA bioprinting systems in combination with DMD array (Fig. 2a) [21, 55, 56]. This preference can be attributed to high popularity of UV-based Type-1 photoinitiators. These systems have been used to demonstrate bioprinting of porous cell-laden structures with high resolution (Fig. 2b–i) [21, 53, 56]. Ma et al. developed a biomimetic hepatic tissue model from human induced pluripotent stem cells (iPSCs) by using a DMD and UV-based SLA 3D bioprinting system [55]. Recently, Miri et al. combined microfluidics-based approach to develop a SLA bioprinting system capable of handling multiple materials [57]. Their system also utilized a DMD and UV-based pattern projection system for cross-linking of hydrogels [21]. However, there is a growing interest toward the use of visible light in SLA bioprinting systems.

A low-cost and high-resolution SLA 3D bioprinting system was developed using a commercially available DLP projector with visible light source and demonstrated comparable biocompatibility to the UV light-based systems [18]. Wang et al. further developed the system to print natural gelatin-based hydrogels demonstrating NIH-3T3 fibroblasts adhesion [57]. Extending the capabilities of SLA bioprinting, Shanjani et al. presented a hybrid 3D bioprinting system utilizing advantages offered by extrusion and SLA bioprinting methods to fabricate multimaterial porous scaffolds [54]. These recent advances in the development of the SLA 3D bioprinting systems highlight the potential and versatility of the SLA bioprinting. An elaborate discussion on the applications of the SLA bioprinting is presented in later sections.

3.2 Direct Laser Writing Bioprinting

Two-photon polymerization is a direct laser writing based high-resolution bioprinting method. Although 2PP cannot be directly classified as SLA bioprinting process, it is a prominent photopolymerization process for printing of microscale ultrafine features. Instead of fabricating 3D scaffold layer by layer, 2PP allows for direct writing within the volume of a photocrosslinkable hydrogel. The apparatus consists of a pulsed laser source (femtosecond laser) and optical components to focus the laser beam. The focused beam results in a highly localized polymerization in the hydrogel volume. 2PP is used for excitation of a photoinitiator using two photons of lower energy instead of a single photon of higher energy, therefore, bypassing the requirement that photoinitiator absorption spectrum should overlap with the wavelength of illumination used. Using the

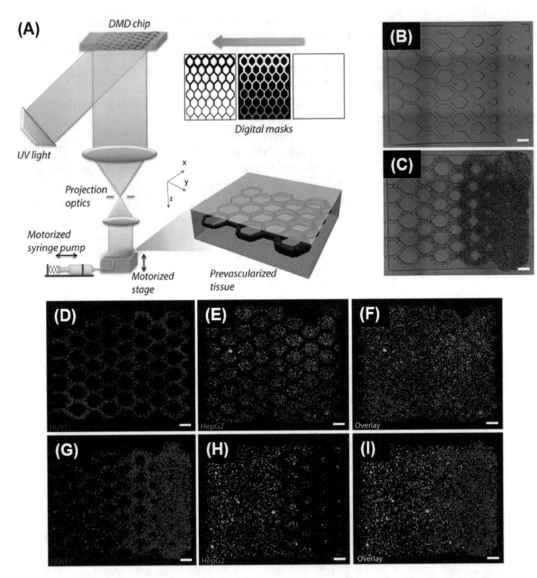

Fig. 2 (a) Schematic of the bioprinting platform. (b) Bioprinted acellular construct featuring intended channels with gradient widths. (c) Bioprinted cellular construct with HUVECs and 10T1/2 (50:1) encapsulated in the intended channels. (d-f) Fluorescent images demonstrating the bioprinting of heterogeneous cell-laden tissue constructs with uniform channel width. HUVECs (red) are encapsulated in the intended channels and HepG2 (green) are encapsulated in the surrounding area. (g-i) Fluorescent images demonstrating the bioprinting of heterogeneous cell-laden tissue constructs with gradient channel widths. Scale bars, 250 μm. (Reprinted from Biomaterials, Vol. 124, Zhu, Wei, et al., Direct 3D bioprinting of prevascularized tissue constructs with complex microarchitecture, 106–115., Copyright (2017), with permission from Elsevier)

pulse width modulation, very high peak intensity (order of tera-watts per cm^2) is obtained such that it overcomes the threshold to initiate two-photon absorption (TPA) [58]. With very high intensity, near-infrared wavelength can be used to cross-link hydrogel responsive to UV range [59, 60].

Ovsianikov et al. have demonstrated the capability of 2PP bioprinting method by fabricating scaffolds with suspended features supported on thinner structures [15, 61]. In a more recent study, Serien and Takeuchi used two-photon direct laser writing to fabricate free-standing 3D microstructure with sub-micron features [62]. The high peak intensity generated by the laser has been considered a limitation in fabrication of cell-laden structure. However, in a recent study, Ovsianikov et al. demonstrated fabrication of cell-containing hydrogels. They also concluded that cell-damage is prominently caused by reactive chemical species rather than laser radiation [63]. Compared to DLP-based SLA bioprinting processes, 2PP bioprinting requires sophisticated apparatus leading to high equipment cost. Being highly localized, the photocrosslinking also requires precise focusing of the laser beam at multiple points in the hydrogel volume which, in turn, slows down the fabrication process but this delay is compensated by extremely short cross-linking times.

4 Applications of SLA 3D Bioprinting

The ability to fabricate complex structures with high resolution has attracted the use of SLA bioprinting in many applications ranging from implantable scaffolds to advanced tissue models on a chip [19, 46, 64, 65]. The specific requirements of biomaterials, cell types, and printing system are defined by the intended applications. Trends in previous researches show that UV-based SLA bioprinting systems are used most widely followed by near-UV wavelength, visible light, and 2PP based systems. As discussed in earlier sections, the popularity of UV-based bioprinting systems can be attributed to fast cross-linking time and high efficiency of the Type-I photoinitiators used to generate free radicals.

Tissue engineering for bone and cartilage has utilized various 3D bioprinting methods [66–68]. Conventional 3D bioprinting methods, however, are limited in their capability of the rapid fabrication of complex structures required for bone scaffolds. The SLA 3D bioprinting relies on the accuracy of the projected masks which can be derived from the CT images of the recipient's bone to rapidly bioprint the bone structure with high precision [69]. Chu et al. fabricated hydroxyapatite implants and reported bone tissue regeneration by controlling the scaffold architecture [70]. Applicability of the SLA bioprinting for 3D cartilage scaffold fabrication was presented by Lee et al., demonstrating high chondrocyte adhesion on scaffold guided by geometry (Fig. 3) [49]. The frequency of occurrence and severity of nervous system injuries has been a major concern and brought about several advancements in the repair of peripheral nervous system (PNS) injuries. Zhu et al.

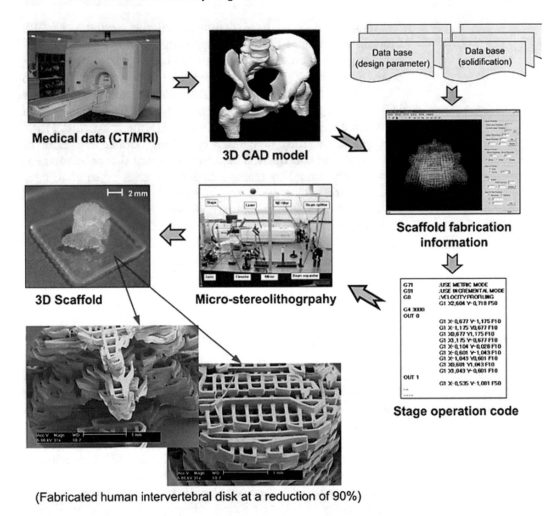

Medical data (CT/MRI)

3D CAD model

Data base (design parameter)

Data base (solidification)

Scaffold fabrication information

3D Scaffold

Micro-stereolithogrpahy

Stage operation code

(Fabricated human intervertebral disk at a reduction of 90%)

Fig. 3 Process flow of SLA bioprinting of 3D cartilage regeneration scaffold. (Reprinted by permission from Springer Nature: Springer Biomedical Microdevices, Application of microstereolithography in the development of three-dimensional cartilage regeneration scaffolds, Seung-Jae Lee, Hyun-Wook Kang, Jung Kyu Park et al., Copyright (2007))

reported UV-based microscale SLA bioprinting to fabricate scaffolds with vascularization networks to aid in peripheral nerve repair [71].

Apart from implantable scaffolds, the SLA bioprinting has also been used to develop models for studying the cell behavior in 3D microenvironments. Ma et al. fabricated a liver module structure with a spatial arrangement of human induced pluripotent stem cells and hepatic progenitor cells for possible application in drug screening (Fig. 4) [55]. Further, the behavior of cancer cells and healthy cells in 3D has been studied to understand the tumorigenic factors. Zhou et al. used a table-top SLA system to bioprint scaffolds with breast cancer cells and bone stromal cells to investigate their interaction and progression of postmetastatic breast cancer in bone

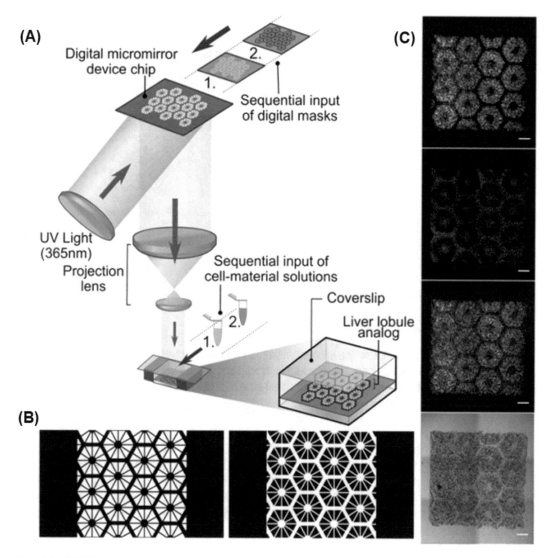

Fig. 4 (a) SLA bioprinting setup schematic for two-step 3D bioprinting. (b) Grayscale digital masks for each bioprinting step. The white patterns represent the light reflecting patterns for photopolymerization. (c) Images (5×) taken under fluorescent and bright field channels showing patterns of fluorescently labeled hiPSC-HPCs (green) in 5% (wt/vol) GelMA and supporting cells (red) in 2.5% (wt/vol) GelMA with 1% GMHA on day 0. Scale bars = 500 μm. (Adapted from Ma et al., *Proceedings of the National Academy of Sciences*, 2016)

[72]. The effect of topographical confinement was reported on endothelial-to-mesenchymal transition by fabricating scaffolds with UV-based physical mask SLA bioprinting [73]. With the ability to fabricate complex 3D scaffolds mimicking the ECM in body, some researches have also studied and identified specific factors guiding the cell differentiation.

In a hybrid bioprinting method, Lee et al. fabricated scaffolds with table-top SLA and electrospinning system, showing high neural stem cell adhesion and neurite extension [48]. Kim et al. also

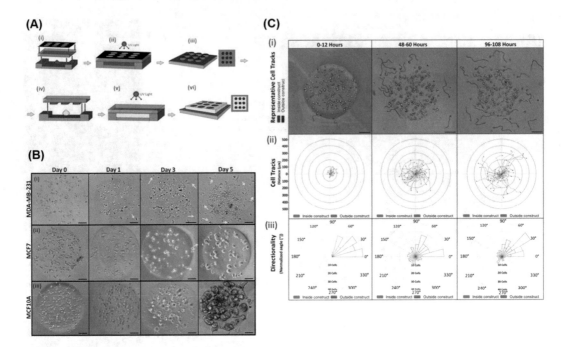

Fig. 5 (**a**) Representative schematic of the microengineered tumor model preparation (steps (*i*)–(*vi*)). Cell laden GelMA is cross-linked first as cylindrical posts with high stiffness, followed by cross-linking of surrounding matrix of low stiffness. (**b**) Representative phase contrast images demonstrating changes in cellular morphology. MDA-MB-231 cells (*i*) spread rapidly creating a heterogeneous (spindle vs. round) morphology. Arrows point to cells that have invaded the surrounding stroma. MCF7 cells (*ii*) exhibited a tendency to cluster, demonstrating only weak migration on days 1 and 3 of culture and small clusters by day 5. MCF10A cells (*iii*) formed similar clusters by day 3 which grew bigger by day 5. Scale bars = 100 μm. (**c**) Representative cell tracks (*i*) of MDA-MB-231 cells within the model. Blue lines indicate tracked cells that are initially within the circular construct at the start of the 12-h period, whereas orange tracks indicate cells that are initially outside the circular construct at the start of the 12-h period. Reconstructed cell tracks (*ii*) normalized with respect to the origin, and 360° rose plots (*iii*) measuring angular directionality. Scale bars = 100 μm. *n* = 156 cells. (Reprinted from Biomaterials, Vol. 81, Peela et al., A three dimensional micropatterned tumor model for breast cancer cell migration studies, 72–83, Copyright (2016), with permission from Elsevier)

presented a review of design parameters affecting osteogenic differentiation in SLA bone scaffold fabrication. Following the research on stem cell differentiation, the design of scaffolds was explored to lead the augmentation of the microenvironment and the manipulation of cell growth and differentiation. In a recent study, Peela et al. demonstrated highly invasive nature of MDA-MB-231 breast cancer cells (Fig. 5) [74]. Using UV-based mask SLA bioprinting, cell encapsulated hydrogel with cylindrical posts of high mechanical stiffness was fabricated. Compared to MCF-7, MDA-MB-231 cells exhibited highly invasive nature by crossing the mechanical stiffness boundary.

5   Concluding Remarks

It is evident that SLA 3D bioprinting has emerged as a promising technology for tissue and organ regeneration. High flexibility and resolution combined with rapid fabrication make it an attractive bioprinting method when compared to other conventional methods. The SLA 3D bioprinting systems in combination with novel biomaterials have been used to develop tissue constructs and study cell behavior and interaction in 3D microenvironments. The research has also diversified in the past decade and more research is being pursued on the development of modular tissues, organs, and disease models in microscale to study various diseases and discover new drugs. The optimization and extensive characterization of the SLA 3D bioprinting systems, on the other hand, have also attracted attention in recent years. In order to address the limitation of the SLA 3D bioprinting in terms of using multiple bioinks, more predictability, and control over photocrosslinking process and resolution, more research toward optimization and improvement in the SLA technology is required in the near future. Utilizing the SLA 3D bioprinting to fabricate highly complex and functional scaffolds with the tuned spatiotemporal behavior is expected to be another prominent research theme in the time ahead.

References

1. O'Connell CD, Zhang B, Onofrillo C et al (2018) Tailoring the mechanical properties of gelatin methacryloyl hydrogels through manipulation of the photocrosslinking conditions. Soft Matter 14:2142–2151

2. Colosi C, Shin SR, Manoharan V et al (2016) Microfluidic bioprinting of heterogeneous 3D tissue constructs using low-viscosity bioink. Adv Mater 28:677–684

3. Koch L, Deiwick A, Chichkov B (2017) Laser additive printing of cells. In: Laser additive manufacturing. Elsevier, Amsterdam, pp 421–437

4. Park JA, Yoon S, Kwon J et al (2017) Freeform micropatterning of living cells into cell culture medium using direct inkjet printing. Sci Rep 7:14610

5. McAllister TN, Maruszewski M, Garrido SA et al (2009) Effectiveness of haemodialysis access with an autologous tissue-engineered vascular graft: a multicentre cohort study. Lancet 373:1440–1446

6. L'heureux N, Pâquet S, Labbé R et al (1998) A completely biological tissue-engineered human blood vessel. FASEB J 12:47–56

7. Lin C-C, Anseth KS (2009) PEG hydrogels for the controlled release of biomolecules in regenerative medicine. Pharm Res 26:631–643

8. Shirahama H, Lee BH, Tan LP, Cho NJ (2016) Precise tuning of facile one-pot gelatin methacryloyl (GelMA) synthesis. Sci Rep 6:1–11. https://doi.org/10.1038/srep31036

9. Yue K, Trujillo-de Santiago G, Alvarez MM et al (2015) Synthesis, properties, and biomedical applications of gelatin methacryloyl (GelMA) hydrogels. Biomaterials 73:254–271. https://doi.org/10.1016/j.biomaterials.2015.08.045

10. Klotz BJ, Gawlitta D, Rosenberg AJWP et al (2016) Gelatin-Methacryloyl hydrogels: towards biofabrication-based tissue repair. Trends Biotechnol 34:394–407. https://doi.org/10.1016/J.TIBTECH.2016.01.002

11. Wang Z, Kumar H, Tian Z et al (2018) Visible light photoinitiation of cell-adhesive gelatin methacryloyl hydrogels for stereolithography

3D bioprinting. ACS Appl Mater Interfaces 10:26859–26869

12. Nichol JW, Koshy ST, Bae H et al (2010) Cell-laden microengineered gelatin methacrylate hydrogels. Biomaterials 31:5536–5544. https://doi.org/10.1016/j.biomaterials. 2010.03.064

13. Lim KS, Schon BS, Mekhileri NV et al (2016) New visible-light Photoinitiating system for improved print Fidelity in gelatin-based bioinks. ACS Biomater Sci Eng 2:1752–1762. https://doi.org/10.1021/acsbiomaterials. 6b00149

14. Noshadi I, Hong S, Sullivan KE et al (2017) In vitro and in vivo analysis of visible light cross-linkable gelatin methacryloyl (GelMA) hydrogels. Biomater Sci 5:2093–2105. https://doi.org/10.1039/C7BM00110J

15. Ovsianikov A, Deiwick A, Van Vlierberghe S et al (2010) Laser fabrication of 3D gelatin scaffolds for the generation of bioartificial tissues. Materials (Basel) 4:288–299. https://doi.org/10.3390/ma4010288

16. Chen Y, Lin R, Qi H et al (2012) Functional human vascular network generated in photo-crosslinkable gelatin methacrylate hydrogels. Adv Funct Mater 22(10):2027–2039. https://doi.org/10.1002/adfm.201101662

17. Samanipour R, Wang Z, Ahmadi A, Kim K (2016) Experimental and computational study of microfluidic flow-focusing generation of gelatin methacrylate hydrogel droplets. J Appl Polym Sci 133:24–26. https://doi.org/10.1002/app.43701

18. Wang Z, Abdulla R, Parker B et al (2015) A simple and high-resolution stereolithography-based 3D bioprinting system using visible light crosslinkable bioinks. Biofabrication 7:045009. https://doi.org/10.1088/1758-5090/7/4/045009

19. Raman R, Bashir R (2015) Chapter 6: Stereolithographic 3D bioprinting for biomedical applications. Elsevier Inc., Amsterdam

20. Soman P, Chung PH, Zhang AP, Chen S (2013) Digital microfabrication of user-defined 3D microstructures in cell-laden hydrogels. Biotechnol Bioeng 110:3038–3047

21. Gauvin R, Chen YC, Lee JW et al (2012) Microfabrication of complex porous tissue engineering scaffolds using 3D projection stereolithography. Biomaterials 33:3824–3834. https://doi.org/10.1016/j.biomaterials.2012.01.048

22. Lin C, Ki CS, Shih H (2015) Thiol–norbornene photoclick hydrogels for tissue engineering applications. J Appl Polym Sci 132:pii: 41563

23. Greene T, Lin TY, Andrisani OM, Lin CC (2017) Comparative study of visible light polymerized gelatin hydrogels for 3D culture of hepatic progenitor cells. J Appl Polym Sci 134:1–10. https://doi.org/10.1002/app. 44585

24. Hoyle CE, Bowman CN (2010) Thiol–ene click chemistry. Angew Chem Int Ed 49:1540–1573

25. Kharkar PM, Rehmann MS, Skeens KM et al (2016) Thiol–ene click hydrogels for therapeutic delivery. ACS Biomater Sci Eng 2:165–179

26. Hermanson GT (2013) Bioconjugate techniques. Academic Press, New York

27. Lutolf MP, Lauer-Fields JL, Schmoekel HG et al (2003) Synthetic matrix metalloproteinase-sensitive hydrogels for the conduction of tissue regeneration: engineering cell-invasion characteristics. Proc Natl Acad Sci 100:5413–5418

28. Lutolf MP, Weber FE, Schmoekel HG et al (2003) Repair of bone defects using synthetic mimetics of collagenous extracellular matrices. Nat Biotechnol 21:513

29. Peng G, Wang J, Yang F et al (2013) In situ formation of biodegradable dextran-based hydrogel via Michael addition. J Appl Polym Sci 127:577–584

30. McGann CL, Levenson EA, Kiick KL (2013) Resilin-based hybrid hydrogels for cardiovascular tissue engineering. Macromol Chem Phys 214:203–213

31. Hao Y, Shih H, Muňoz Z et al (2014) Visible light cured thiol-vinyl hydrogels with tunable degradation for 3D cell culture. Acta Biomater 10:104–114

32. Shih H, Lin C (2013) Visible-light-mediated thiol-ene hydrogelation using eosin-Y as the only photoinitiator. Macromol Rapid Commun 34:269–273

33. Qin X-H, Ovsianikov A, Stampfl J, Liska R (2014) Additive manufacturing of photosensitive hydrogels for tissue engineering applications. BioNanoMaterials 15:49–70. https://doi.org/10.1515/bnm-2014-0008

34. Williams CG, Malik AN, Kim TK et al (2005) Variable cytocompatibility of six cell lines with photoinitiators used for polymerizing hydrogels and cell encapsulation. Biomaterials 26:1211–1218. https://doi.org/10.1016/j.biomaterials.2004.04.024

35. Fairbanks BD, Schwartz MP, Bowman CN, Anseth KS (2009) Photoinitiated polymerization of PEG-diacrylate with lithium phenyl-2,4,6-trimethylbenzoylphosphinate: polymerization rate and cytocompatibility. Biomaterials

30:6702–6707. https://doi.org/10.1016/j.biomaterials.2009.08.055

36. Ullrich G, Burtscher P, Salz U et al (2006) Phenylglycine derivatives as coinitiators for the radical photopolymerization of acidic aqueous formulations. J Polym Sci Part A Polym Chem 44:115–125. https://doi.org/10.1002/pola.21139

37. Bahney CS, Lujan TJ, Hsu CW et al (2011) Visible light photoinitiation of mesenchymal stem cell-laden bioresponsive hydrogels. Eur Cells Mater 22:43–55

38. Popielarz R, Vogt O (2008) Effect of coinitiator type on initiation efficiency of two-component photoinitiator systems based on eosin. J Polym Sci Part A Polym Chem 46:3519–3532

39. Li L, Fourkas JT (2007) Multiphoton polymerization. Mater Today 10:30–37

40. Bártolo PJ (2011) Stereolithographic processes. In: Stereolithography. Springer, New York, pp 1–36

41. Zhou C, Chen Y, Waltz RA (2009) Optimized mask image projection for solid freeform fabrication. J Manuf Sci Eng 131:061004. https://doi.org/10.1115/1.4000416

42. Sun C, Fang N, Wu DM, Zhang X (2005) Projection micro-stereolithography using digital micro-mirror dynamic mask. Sensors Actuators A Phys 121:113–120. https://doi.org/10.1016/j.sna.2004.12.011

43. Larson C, Shepherd R (2016) 3D bioprinting technologies for cellular engineering. In: Microscale Technologies for Cell Engineering. Springer, New York, pp 69–89

44. Hornbeck LJ (1996) Multi-level digital micro-mirror device

45. Arcaute K, Mann BK, Wicker RB (2006) Stereolithography of three-dimensional bioactive poly(ethylene glycol) constructs with encapsulated cells. Ann Biomed Eng 34:1429–1441. https://doi.org/10.1007/s10439-006-9156-y

46. Chan V, Zorlutuna P, Jeong JH et al (2010) Three-dimensional photopatterning of hydrogels using stereolithography for long-term cell encapsulation. Lab Chip 10:2062. https://doi.org/10.1039/c004285d

47. Zorlutuna P, Jeong JH, Kong H, Bashir R (2011) Stereolithography-based hydrogel microenvironments to examine cellular interactions. Adv Funct Mater 21:3642–3651. https://doi.org/10.1002/adfm.201101023

48. Lee S-J, Nowicki M, Harris B, Zhang LG (2017) Fabrication of a highly aligned neural scaffold via a table top stereolithography 3D printing and electrospinning. Tissue Eng Part A 23:491–502. https://doi.org/10.1089/ten.tea.2016.0353

49. Lee SJ, Kang HW, Park JK et al (2008) Application of microstereolithography in the development of three-dimensional cartilage regeneration scaffolds. Biomed Microdevices 10:233–241. https://doi.org/10.1007/s10544-007-9129-4

50. Knowlton S, Anand S, Shah T, Tasoglu S (2017) Bioprinting for neural tissue engineering. Trends Neurosci 41:31. https://doi.org/10.1016/j.tins.2017.11.001

51. Wüst S, Müller R, Hofmann S (2011) Controlled positioning of cells in biomaterials—approaches towards 3D tissue printing. J Funct Biomater 2(3):119–154

52. Lu Y, Mapili G, Suhali G et al (2006) A digital micro-mirror device-based system for the microfabrication of complex, spatially patterned tissue engineering scaffolds. J Biomed Mater Res: Part A 77:396–405. https://doi.org/10.1002/jbm.a.30601

53. Mapili G, Lu Y, Chen S, Roy K (2005) Laser-layered microfabrication of spatially patterned functionalized tissue-engineering scaffolds. J Biomed Mater Res: Part B Appl Biomater 75:414–424. https://doi.org/10.1002/jbm.b.30325

54. Shanjani Y, Pan CC, Elomaa L, Yang Y (2015) A novel bioprinting method and system for forming hybrid tissue engineering constructs. Biofabrication 7:45008. https://doi.org/10.1088/1758-5090/7/4/045008

55. Ma X, Qu X, Zhu W et al (2016) Deterministically patterned biomimetic human iPSC-derived hepatic model via rapid 3D bioprinting. Proc Natl Acad Sci 113:2206–2211

56. Zhang AP, Qu X, Soman P et al (2012) Rapid fabrication of complex 3D extracellular microenvironments by dynamic optical projection stereolithography. Adv Mater 24:4266–4270. https://doi.org/10.1002/adma.201202024

57. Miri AK, Nieto D, Iglesias L et al (2018) Microfluidics-enabled multimaterial maskless stereolithographic bioprinting. Adv Mater 30 (27):e1800242

58. Nguyen AK, Narayan RJ (2017) Two-photon polymerization for biological applications. Mater Today 20:314

59. Lee K-S, Kim RH, Yang D-Y, Park SH (2008) Advances in 3D nano/microfabrication using two-photon initiated polymerization. Prog Polym Sci 33:631–681

60. Maruo S, Nakamura O, Kawata S (1997) Three-dimensional microfabrication with two-photon-absorbed photopolymerization. Opt Lett 22:132–134

61. Ovsianikov A, Chichkov BN (2012) Three-dimensional microfabrication by two-photon polymerization technique. In: Computer-aided tissue engineering. Springer, New York, pp 311–325

62. Serien D, Takeuchi S (2017) Multi-component microscaffold with 3D spatially defined proteinaceous environment. ACS Biomater Sci Eng 3:487–494. https://doi.org/10.1021/acsbiomaterials.6b00695

63. Ovsianikov A, Mühleder S, Torgersen J et al (2013) Laser photofabrication of cell-containing hydrogel constructs. Langmuir 30:3787–3794

64. Warner J, Soman P, Zhu W et al (2016) Design and 3D printing of hydrogel scaffolds with fractal geometries. ACS Biomater Sci Eng 2:1763–1770. https://doi.org/10.1021/acsbiomaterials.6b00140

65. Bens A, Seitz H, Bermes G et al (2007) Non-toxic flexible photopolymers for medical stereolithography technology. Rapid Prototyp J 13:38–47. https://doi.org/10.1108/13552540710719208

66. Huang Y, Zhang X-F, Gao G et al (2017) 3D bioprinting and the current applications in tissue engineering. Biotechnol J 2017:1600734. https://doi.org/10.1002/biot.201600734

67. Hutmacher DW (2000) Scaffolds in tissue engineering bone and cartilage. Biomaterials 21:2529–2543

68. Bose S, Vahabzadeh S, Bandyopadhyay A (2013) Bone tissue engineering using 3D printing. Mater Today 16:496–504

69. Izatt MT, Thorpe PLPJ, Thompson RG et al (2007) The use of physical biomodelling in complex spinal surgery. Eur Spine J 16:1507–1518

70. Chu TMG, Orton DG, Hollister SJ et al (2002) Mechanical and in vivo performance of hydroxyapatite implants with controlled architectures. Biomaterials 23:1283–1293. https://doi.org/10.1016/S0142-9612(01)00243-5

71. Zhu W, Qu X, Zhu J et al (2017) Direct 3D bioprinting of prevascularized tissue constructs with complex microarchitecture. Biomaterials 124:106–115. https://doi.org/10.1016/j.biomaterials.2017.01.042

72. Zhou X, Zhu W, Nowicki M et al (2016) 3D bioprinting a cell-laden bone matrix for breast Cancer metastasis study. ACS Appl Mater Interfaces 8:30017–30026. https://doi.org/10.1021/acsami.6b10673

73. Nasrollahi S, Pathak A (2016) Topographic confinement of epithelial clusters induces epithelial-to-mesenchymal transition in compliant matrices. Sci Rep 6:1–12. https://doi.org/10.1038/srep18831

74. Peela N, Sam FS, Christenson W et al (2016) A three dimensional micropatterned tumor model for breast cancer cell migration studies. Biomaterials 81:72–83. https://doi.org/10.1016/j.biomaterials.2015.11.039

Part II

Protocols for 3D Bioprinting

Chapter 7

Characterizing Bioinks for Extrusion Bioprinting: Printability and Rheology

Cathal O'Connell, Junxiang Ren, Leon Pope, Yifan Zhang, Anushree Mohandas, Romane Blanchard, Serena Duchi, and Carmine Onofrillo

Abstract

In recent years, new technologies based on 3D bioprinting have emerged as ideal tools with which to arrange cells and biomaterials in three dimensions and so achieve tissue engineering's original goals. The simplest and most widely used form of bioprinting is based on pneumatic extrusion, where 3D structures are built up by drawing patterns of cell-laden or non–cell-laden material through a robotically manipulated syringe. Developing and characterizing new biomaterials for 3D bioprinting (i.e., bioinks) is critical for the progress of the field. This chapter describes a series of protocols for developing, optimizing, and testing new bioinks for extrusion-based 3D bioprinting.

Key words Bioink, 3D Bioprinting, Biofabrication, Rheology, Printability, Compressive modulus

1 Introduction

Tissue engineering was defined by Langer and Vacanti as a field which applies the principles of biology and engineering to the development of functional substitutes for damaged tissue [1]. In recent years, new technologies based on 3D bioprinting have emerged as ideal tools with which to arrange cells and biomaterials in three dimensions, and so achieve tissue engineering's original goals [2]. The simplest and most widely used form of bioprinting is based on pneumatic extrusion, where 3D structures are built up by drawing patterns of cell-laden or non–cell-laden material through a robotically manipulated syringe.

The material extruded through a bioprinter is called the "bioink" [3]. To maintain cell viability the bioink is typically a

---

The original version of this chapter was revised. The correction to this chapter is available at https://doi.org/10.1007/978-1-0716-0520-2_18

Jeremy M. Crook (ed.), *3D Bioprinting: Principles and Protocols*, Methods in Molecular Biology, vol. 2140, https://doi.org/10.1007/978-1-0716-0520-2_7, © Springer Science+Business Media, LLC, part of Springer Nature 2020, Corrected Publication 2022

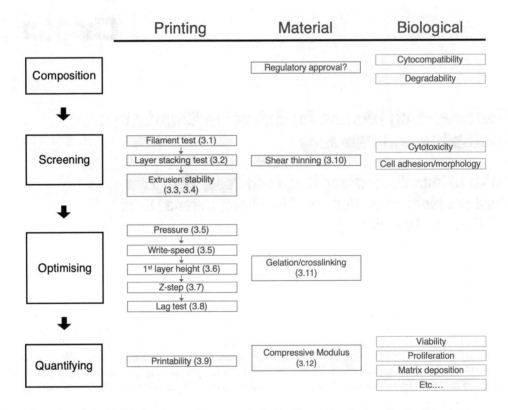

Fig. 1 Overview of the bioink development process, including the protocols described in this chapter

hydrogel. A plethora of hydrogel materials have been developed, with extremely high water content, porosity, and permeability, all designed to mimic the extracellular environment [4]. The bioink may be designed to play only a temporary role, merely as a sacrificial support structure [5, 6]. Alternatively, the bioink may be designed to form a 3D environment critical to the attachment and differentiation of cells within the substitute tissue. However, cell phenotype and differentiation is strongly mediated by both their mechanical and biochemical microenvironment [7–9].Thus, the first step for a new bioprinting based research strategy is typically the development of a new bioink, tailored to the specific tissue being targeted. This chapter describes a series of protocols for developing and testing new bioinks for extrusion-based 3D bioprinting (Fig. 1).

1.1 Biological Requirements

The first step in any bioink development program is to consider the biological environment and requirements of the target tissue. These requirements will determine the palette of materials from which the bioink can be formulated. The biochemical microenvironments of the desired tissue should be considered, in particular the relevance of cell-attachment sites, as these can trigger cell differentiation. Degradability of the material must be considered, as well as the potential toxicity of degradation products. For work targeting clinical translation, the regulatory approval status of all components as well as the sources of the specific materials used in any trials should be assessed as early as possible [10]. Where possible initial screening for compatibility of the candidate materials with the target application (for example, cell differentiation testing) should be performed prior to further ink development.

1.2 Materials Screening

Extrusion based bioprinting is a relatively forgiving printing mode, amenable to inks with a wide range of viscosities and surface tensions. Nevertheless, there are constraints on what materials can form free-standing 3D structures. A good proxy measure of printability is the formation of a filament (also called a "string") morphology when extruded from a nozzle, rather than a droplet [11]. Subheading 3.1.1 below describes a simple test for filament morphology. Meanwhile, to form self-supporting 3D structures; the bioink must be able to form multiple stacked layers, without fusing [12]. Subheading 3.1.2 describes a simple test for layer stacking. Only materials which pass both of these tests should be considered for further development.

For consistent, high quality printing, the bioink must exhibit stable and reproducible printing properties. Subheading 3.2.1 describes a method for measuring volumetric extrusion rate [13], and can be performed repeatedly over time to assess stability, or, for example, as a function of extrusion pressure or temperature. Subheading 3.2.2 describes a method for measuring extrusion force. This method is especially useful for quantifying the homogeneity of a formulation; varying forces indicate the material is inhomogeneous, while spiking of the force is the hallmark of a clogged nozzle.

1.3 Printing Optimization

Once the bioink is formulated, it is important to optimize the printing conditions as a means to maximize both resolution and repeatability. Several interrelated parameters must be optimized, including the extrusion pressure, feed rate (also called the write speed), and the z-stepping height between layers. We advocate a sequential optimization process, beginning with the choice of an extrusion pressure that achieves a slow (~5 mm/s) but steady filament. For many hydrogel materials, stable printing is achieved through a minimization of net lateral forces on the extruded filament. This means matching the speed of movement of the nozzle, to that of the extrusion rate of the filament. Subheading 3.3.1 describes a protocol for measuring the filament extrusion speed at a given pressure. Typically, filament diameter is determined by nozzle gauge, and does not depend on extrusion pressure.

The distance between the nozzle and substrate during printing is a critical parameter, and one of the trickiest to optimize in 3D bioprinting. Setting this distance too high can result in an unrecognizable blob, while setting the distance too low can impact on printing resolution by squashing the deposited filament, and increasing line-width. The z-stepping distance is a function of the nozzle diameter, as well as any deformation of the layers due to gel–substrate interactions and/or gravity. As a rule of thumb, the z-

stepping distance is nominally about ½ the nozzle diameter. The first layer (gel on substrate) is typically different from all subsequent layers (gel on gel) and so must be optimized separately. Subheadings 3.3.2 and 3.3.3 describe protocols for optimizing the first layer height and subsequent z-stepping heights respectively.

Some bioprinting systems will exhibit a significant lag time between the onset of the extrusion pressure, and the actual extrusion of material. Conversely, there can be a lag time between the switching off of the extrusion pressure, and the cessation of actual material extrusion. These log on and lag off timings can be detrimental to the printed pattern in various ways, and should be accounted for in the programming of bioprinter. Subheading 3.3.4 describes a simple method to quantify these lag times.

To assess the relative merits of candidate bioinks, or printing parameters, it is useful to have a quantitative measure of printability. Subheading 3.3.5 describes a method for quantifying the fidelity of a printed pattern using image analysis.

1.4 Bioink Rheology

To understand the printing performance of a bioink, it is critical to study its rheological properties. For example, successful bioinks typically exhibit a drop in viscosity during flow (shear-thinning behavior), as this effectively reduces the shear stresses experienced by encapsulated cells as they extrude through the nozzle. Subheading 3.4.1 describes a method for quantifying this shear-thinning behavior by measuring the viscosity of a bioink as a function of shear rate.

Rheology can also be used to study gelation of the bioink, for example, as a function of temperature or exposure to a cross-linking agent. Subheading 3.4.2 describes a method for following the gelation process by probing the viscoelastic properties of a bioink using oscillating shear rheology.

1.5 Mechanical Testing

For many tissues, particularly for load-bearing tissues such as bone or cartilage, the mechanical properties of the printed structure are critical. Subheading 3.4.3 below describes a method for performing mechanical testing of hydrogels by unconfined compression.

2 Materials

2.1 Bioprinting (Fig. 2)

1. Pneumatic extrusion bioprinter (for the example data below, a Cellink Inkredible was used).

2. Pneumatic tubing (3 mm inner diameter).

3. Pneumatic connector for bioink cartridge (Nordson EFD).

4. 3 mL pneumatic extrusion cartridge (Nordson EFD).

5. Pneumatic piston (Nordson EFD).

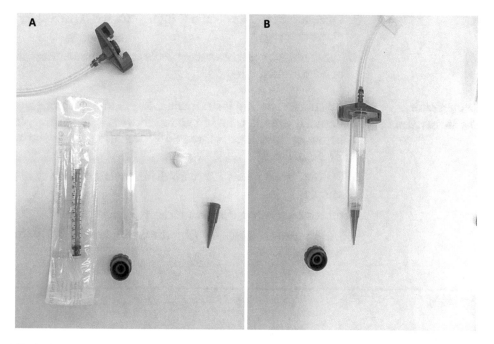

Fig. 2 Typical consumables required for extrusion based bioprinting: pneumatic tubing and connector, 3 mL cartridge, pneumatic piston (white), cartridge cap (blue) to close cartridge when nozzle not attached, 250 μm tapered nozzle (red), 1 mL syringe. Left, disassembled. Right, assembled

6. Cartridge cap (Nordson EFD).

7. Extrusion nozzle (Nordson EFD).

8. Ruler and stopwatch.

9. 1 mL syringe for transferring materials.

10. Bioink to be tested.

11. A small USB stereomicroscope (e.g., DinoLite), which can be mounted on a retort stand.

2.2 Extrusion Force Measurement

1. Mechanical testing instrument incorporating load cell, and data logging components (for the example data below, a Bose Electroforce 5500, equipped with a 22 N load cell).

2. A syringe driver equipped with a "max force" parameter (for the example data below, a Harvard PhD Ultra was used).

3. Bioprinting consumables as described above.

4. Bioink to be tested.

2.3 Rheology

1. Rheometer (for the example data below, an Anton Parr MCR302 was used).

2. Measuring cone (for the example data below, a 15 mm diameter cone with 1° cone angle was used).

3. Quartz base-plate and light source (necessary for photorheology).

4. Temperature control system (necessary for temperature dependent measurements).

2.4 Compressive Modulus Measurement

1. Mechanical testing instrument incorporating load cell and data logging, as well as an integrated mover (for the example data below, a Bose Electroforce 5500 was used, equipped with a 22 N load cell).

2. Stainless steel plates ×2 (with diameter larger than that of the sample).

3. Stereomicroscope (for area measurement).

4. Image analysis software (for the example data, FIJI/ImageJ was used).

3 Methods

3.1 Printability Screening Procedures

3.1.1 String/Filament Test

1. Load bioink material into a 3 mL pneumatic cartridge. For viscous inks, use a 1 mL syringe (without a needle) to transfer the material. A Luer lock female-female coupler can also be used to transfer viscous bioinks between syringes directly (*see* **Note 1**).

2. Assemble the cartridge with a nozzle (e.g., 250 μm diameter tapered nozzle) and pneumatic piston (*see* **Note 2**).

3. Attach pneumatic tubing to the cartridge. Connect the tubing to an air pressure regulator or 3D bioprinter.

4. Grip the cartridge in a vertical position.

5. Focus a suitable camera (e.g., a retort-stand mounted stereomicroscope) on the end of the nozzle.

6. Slowly increase air pressure in 5 kPa increments until the bioink begins to be extruded steadily.

7. Observe the morphology of the bioink as it emerges from the nozzle (Fig. 3) (*see* **Note 3**).

   (a) Formation of a liquid droplet, which drops from the nozzle, indicates a test fail.

   (b) Formation of a thin filament, which hangs >5 mm from the tip of the nozzle, indicates a test pass. The length of the filament before it ruptures can also be used as a relative measure of printability [14].

8. Only bioinks which exhibit string morphology should be considered for subsequent printability tests (*see* **Note 4**).

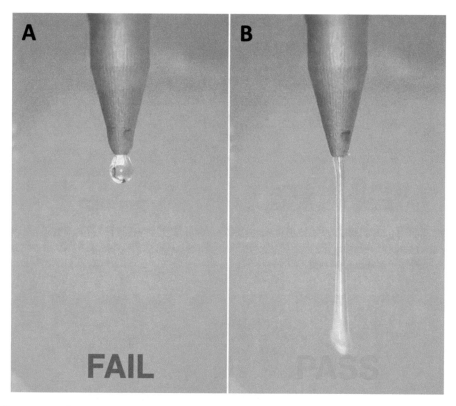

Fig. 3 Filament test. The bioink on the left formed a liquid drop at the end of the nozzle. The bioink on the right formed a filament upon extrusion, indicating a promising material for 3D bioprinting

3.1.2 Layer Stacking Test

1. Load bioink material into a 1 mL syringe and attach a 250 μm diameter nozzle.

2. Manually apply pressure to the syringe plunger to extrude a thin line of material onto a plastic petri dish.

3. With the nozzle raised slightly from the substrate, draw a second line at right angles to the first, and crossing it to make a "+" sign.

4. Observe under a microscope whether the two lines have merged together (fail), or stack atop one another (pass) (Fig. 4).

3.2 Extrusion Stability Procedures

3.2.1 Extrusion Rate Measurement I: Volume

1. Load bioink material into a 3 cm³ pneumatic cartridge.

2. Assemble the cartridge with a nozzle (e.g., 250 μm diameter tapered nozzle) and pneumatic piston.

3. Attach pneumatic tubing to the cartridge. Connect the tubing to an air pressure regulator or 3D bioprinter.

4. Slowly increase air pressure until the bioink begins to be extruded steadily.

5. Preweigh a small sample vial (with airtight cap).

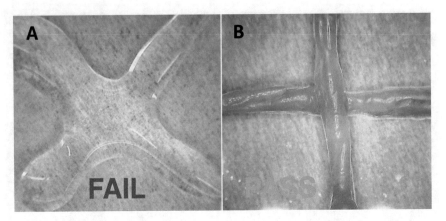

Fig. 4 Layer stacking test. Left, the lines of this liquid-like bioink (25% pluronic F-127 in water) coalesce when printed onto a substrate—failing the test. Right, the lines of this bioink (33% pluronic F-127 in water) stack on top of one another—passing the test

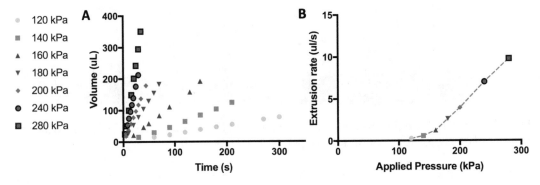

Fig. 5 Volume extrusion rate data for a typical bioink (gelatin methacryloyl 10% w/v, hyaluronic acid methacryloyl 2% w/v) through a cylindrical nozzle of diameter 0.41 mm. Shear thinning behavior (*see* section 3.4.1 below) is evident in the nonlinearity of the extrusion rate as a function of applied pressure

6. Extrude the bioink for a known duration (e.g., 5 s) and capture the extruded material into the preweighed vial.

7. Weigh the mass of the extruded material.

8. Repeat **steps 6** and **7** several times.

9. Convert the mass of extruded material to volume using the density of the bioink (density = mass/volume).

10. Plot volume extruded against the extrusion time (Fig. 5). The slope of this plot is the volume extrusion rate at the chosen pressure.

11. To assess the extrusion stability over time, measuring the extrusion rate at several relevant time points, for example, at $t = 15, 30, 60, 120$ min after printer setup. For consistent printing, a bioink must exhibit stable extrusion rate over the timescale of a printing experiment (Fig. 5) (*see* **Note 5**).

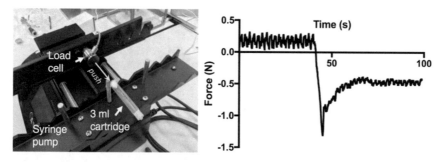

Fig. 6 Left, experimental set up with the pneumatic cartridge and load cell arranged on a syringe pump. Right, example data of extrusion force as a function of time. Here, the load cell makes contact with the piston at the ~40 s mark. The initial spike of force (−1.3 N) is due to static friction of the piston/cartridge. Stable extrusion force of ~0.5 N is then observed

3.2.2 Extrusion Force Measurement

1. Load bioink material into a 3 mL pneumatic cartridge.

2. Assemble the cartridge with a nozzle (e.g., 250 μm diameter tapered nozzle) and pneumatic piston.

3. Place the cartridge in a suitable syringe pump. *WARNING: Use a syringe pump with the ability to set a "Maximum Force." Set the Maximum Force at a level below that which might damage the load cell.*

4. Attach the load cell onto one arm of the syringe pump (*see* Fig. 6). Arrange the apparatus such that the load cell pushes the piston inside the cartridge (e.g., using a small rod affixed to the load cell).

5. On the syringe pump, choose a rate of extrusion that corresponds with your bioprinting experiments. For some syringe pump models, you may need to input the dimensions (i.e., internal diameter) of the cartridge to control the volume extrusion rate.

6. Start the syringe pump. On the datalogger or mechanical testing device, record the force measured by the load cell as a function of time (Fig. 6).

3.3 Bioprinting Optimization (Fig. 7)

3.3.1 Extrusion Rate Measurement II: Filament Speed

1. Load bioink material into a 3 cm^3 pneumatic cartridge.

2. Assemble the cartridge with a nozzle (e.g., 250 μm diameter tapered nozzle) and pneumatic piston.

3. Attach pneumatic tubing to the cartridge. Connect the tubing to an air pressure regulator or 3D bioprinter.

4. Set up a ruler next to the nozzle, held vertically.

5. Situate a stopwatch behind the nozzle.

6. Set the camera and adjust so that nozzle, ruler and stopwatch are all captured in the frame (Fig. 8a). Make sure the nozzle tip is on the top of the frame, this will maximize the distance needed to observe the filament displacement.

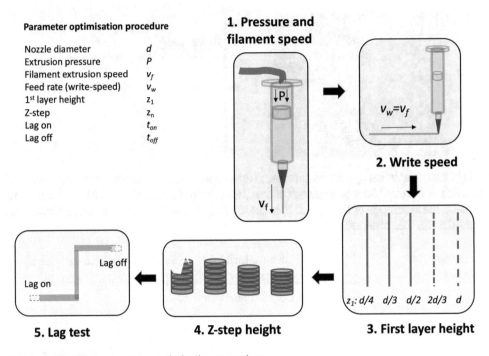

Fig. 7 Schematic of the parameter optimization procedure

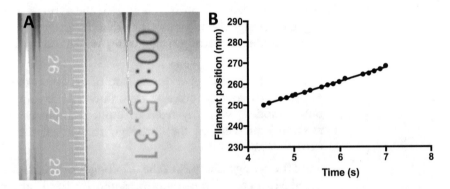

Fig. 8 (a) Set-up for the "Extrusion Rate: Filament Speed" measurement, with a ruler arranged next to the nozzle and a stopwatch visible behind. (b) The position of the filament is plotted against the stopwatch time points. The slope of this line yields the extrusion rate in mm/s. Extrusion speed: 6.645 ± 0.078 mm/s, R^2 0.9987. Data for 35% pluronic F-127 in PBS solution, extruded at a pressure of 46 kPa, through a tapered nozzle of diameter 250 µm

7. Slowly increase air pressure until the bioink begins to be extruded steadily.

8. Start the stopwatch followed by video recording.

9. Switch on the extrusion pressure, and observe the extrusion of the string.

10. On still frames of the video, estimate filament displacement at each time point with reference to the ruler and stopwatch.

11. Plot the filament displacement and time point data (Fig. 8b). The filament extrusion speed is the slope of this line (e.g., in mm/s).

12. To assess the extrusion stability over time, measuring the extrusion rate at several relevant time points, for example, at $t = 15, 30, 60, 120$ min. For consistent printing, a bioink must exhibit stable extrusion rate over the timescale of a printing experiment (*see* **Note 6**).

3.3.2 Bioprinting Parameter Optimization Procedure: First Layer

1. Load bioink material into a pneumatic cartridge. Attach a nozzle (e.g., 250 μm diameter) and pneumatic piston.

2. Set up the cartridge on a 3D bioprinter.

3. Calibrate the z-height of the nozzle as per the bioprinter's instructions.

4. With the printer nozzle far from the surface, turn on the extrusion pressure.

5. Adjust the pressure until the extruded filament is moving at approximately 10 mm/s. (*For more accuracy, use the filament extrusion rate measurement protocol.*)

6. Define the feed rate (also called write speed) of the printer as 10 mm/s (*or the measured rate*).

7. Define a printing pattern (i.e., using the printer software or using g-code) (*see* **Note 7**) such that five lines are printed at incrementally increasing z-height. For example, for a 250 μm nozzle, choose $z = \{0.05, 0.1, 0.15, 0.2, 0.25\}$ mm.

8. Start the print. Observe the deposition of the successive lines.

9. Use a microscope with a large field of view to capture a record of the printed pattern (Fig. 9).

10. Observe the threshold below which the printed lines are continuous. The first layer z-height is typically ~½ nozzle diameter. Lower z-heights increase adhesion to the substrate, while also increasing line width.

3.3.3 Bioprinting Parameter Optimization Procedure: Z-Step

1. Set up the bioprinter as described previously.

2. Define previously optimized printing pressure, write speed, and first layer height.

3. Define a printing pattern (i.e., using the printer software or using g-code—*see* **Note 7**) such that a series of cylinders are printed with incrementally increasing z-steps. Each cylinder should be ten layers tall.

4. Run the print.

5. Use a horizontally mounted stereomicroscope to image the printed cylinders in profile (Fig. 10). The optimum layer height is marked by the tallest well-defined cylinder.

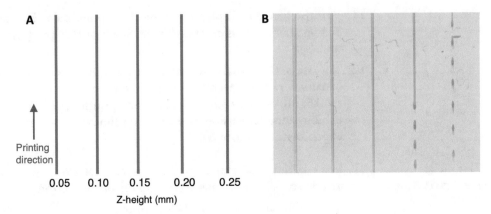

Z-height (mm)

Fig. 9 Left, example pattern design for a first layer height test using a nozzle with 0.25 mm inner diameter. Right, example data using a pluronic F-127 solution. In this case, lines printed with *z*-heights of 0.05, 0.1 and 0.15 mm are acceptable. Lines printed at *z*-heights of 0.2 and 0.25 mm are spotty, and unacceptable. The optimum *z*-height, for consistent printing at highest resolution, was chosen as 0.15 mm

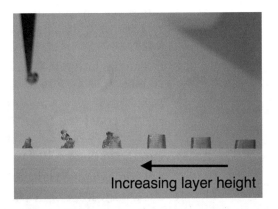

Fig. 10 Example data from *z*-step optimization test. *Z*-step increments increase from right to left. The optimum layer height is the tallest well defined cylinder (i.e., third from the right). The bioprinting nozzle at top left has picked up some bioink during the failed prints

3.3.4 Lag Test

1. Set up the bioprinter as described previously.

2. Define previously optimized printing pressure, write speed, and first layer height.

3. Define a printing pattern (i.e., using the printer software or using g-code (*see* **Note 7**) such that a zigzag pattern (Fig. 11): 10 mm movement in the *x*-direction (A–B), followed by 10 mm in the *y*-direction (B–C), followed by a final 15 mm movement in the *x*-direction (C–E). The extrusion pressure should turn on at the onset of the first *x*-movement (Point A), but turn off during the second *z*-movement (Point D, while the nozzle is still moving toward Point E).

4. Run the print.

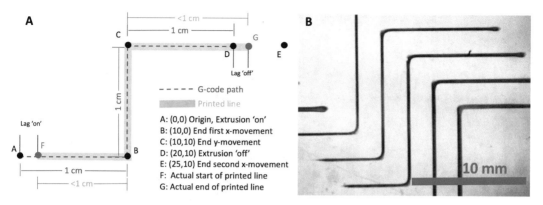

Fig. 11 Lag-time test

5. Capture an image of the printed pattern. Measure the lengths FB, BC, and CG in mm. The lag times can be calculated from the following equations:

$$\frac{\text{Length BC} - \text{Length FB}}{\text{Movement speed}} = \text{Lag}'\text{on}'\text{time}$$

$$\frac{\text{Length CG} - \text{Length BC}}{\text{Movement speed}} = \text{Lag}'\text{off}'\text{time}$$

3.3.5 Quantification of Bioink Printability

1. Set up the bioprinter as described previously.

2. Define previously optimized printing pressure, write speed, layer heights, and lag times.

3. Define a printing pattern (i.e., using the printer software or using g-code (*see* **Note 7**) creating a two-layered cross-hatch structure with line spacing of 1 mm.

4. Run the print.

5. Use a microscope to capture an image showing at least four pores (Fig. 12).

6. Using image analysis software (e.g., ImageJ), measure the internal perimeter (L) and internal area (A) of each pore.

7. The "Printability" is defined as

$$\text{Pr} = \frac{L^2}{16A}$$

where a value of $\text{Pr} = 1$ indicates a perfectly defined square shape. $\text{Pr} < 1$ indicates overly circular pores, typically as a result of low viscosity bioinks. $\text{Pr} > 1$ indicates an irregular shape, typically caused by over-gelation of the bioink.

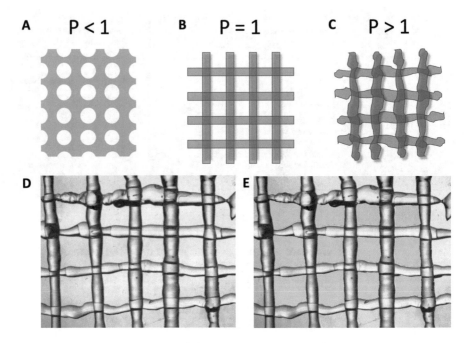

Fig. 12 Top, schematic of three regimes of printability. Pr < 1 is generated from overly circular inner pores, indicative of a liquidy, undergelated bioink. Pr = 1 indicates square pores, meaning well defined lines and good printability. Pr > 1 is generated from wobbly lines, indicating an overly gelled bioink material. Bottom, stereomicroscope images of a two-layer structure printed according to Subheading 3.3.5 (Pr = 1.04, meaning a slightly overly gelled bioink)

3.4 Rheology

3.4.1 Rheology: Shear Rate

1. Set up rheometer according to the equipment's standard operating procedure.

2. Load the measuring system (e.g., cone and plate). For the example data below a stainless steel cone with 15 mm diameter and a 1° cone angle was used (Fig. 13).

3. Choose the option in the rheometer software to reset the normal force (i.e., to tare the weight of the measuring system.

4. Choose the option in the rheometer software to set the zero gap.

5. When this is done, raise the measuring system far from the surface (e.g., 20 mm).

6. If the rheometer has temperature control capability, set the desired temperature for the measurement.

7. In the measuring profile settings, choose a "flow" or "continuous ramp" measurement.

8. Select a continuous shear rate ramp from 0.0001 to 100 s^{-1}.

9. Select the data collection settings appropriate for the rheometer (in the example data shown, measuring points were taken every 2 s for the duration of the measurement; 180 s) (Fig. 13).

A

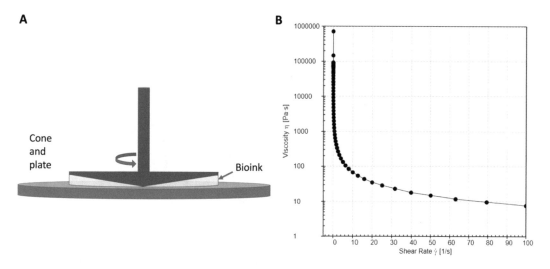

Fig. 13 Left, schematic of continuous cone and plate rheology. The material is held between a flat plate at bottom, and a cone shaped measuring system on top. The measuring cone is rotated by the system at a continually ramped speed, and the torque opposing this rotation is measured. These data are used (usually automatically by the rheometer software) to calculate viscosity as a function of the shear rate. Right, plot of viscosity over shear rate for the shear thinning hydrogel pluronic F-127

10. Check the sample volume required for the measuring cone being used.

11. Dispense about 110% of this volume onto the center of the bottom plate.

12. Lower the measuring cone to its measuring gap. (Measuring cones each have a designated measuring gap given by the height of the slight truncation at the tip. For the example data below, the measuring gap was 31 μm).

13. The rheometer will first move to a trim position. For viscous bioinks, use a flat spatula trim the material from around the edges of the measuring cone. For liquidy inks, do not use a spatula, but instead carefully control the dispensing volume so that the bioink completely fills the measuring gap with no excess. *See* **Note 8**.

14. Continue the movement to the measuring position.

15. Start the measurement.

3.4.2 Rheology: Oscillation

1. Follow **steps 1–6** as described in section 3.4.1.

2. In the measuring profile settings, choose "oscillation" or "viscoelastic."

3. Select constant angular oscillation, with appropriate oscillation frequency and strain (typically used parameters, as in the example date below, are 1% strain at 10 rad/s).

A

B

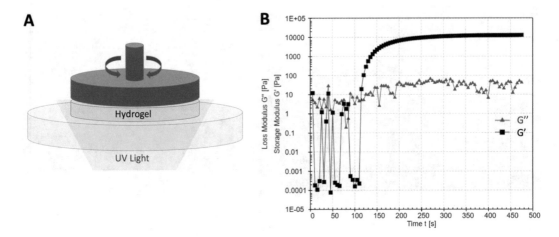

Fig. 14 Left, schematic of oscillatory rheology set up to monitor photocrosslinking in situ. The measuring cone is oscillated at a known rate, and the torque resisting this motion is measured. In viscoelastic materials, the storage modulus (G') measure the stored energy, representing the elastic portion. The loss modulus (G") measures the energy lost as heat, representing the viscous portion. For liquids, the loss modulus dominates, while for solids (or gels) the storage modulus dominates. The materials undergoing a phase transition, the "gelation point" is defined as the point where the storage modulus increases above the loss modulus. Right, example data from in situ photorheology of 10% gelatin methacryloyl including 0.5% Irgacure 2959 photo-initiator and illuminated with 365 nm light at 50 mW/cm². The light was turned on at the 100 s time point. The gelation point is at about the 110 s time point when the storage modulus (black squares) rises above the loss modulus (red triangles). The storage modulus continues to rise until about the 300 s time point, at which the storage modulus reaches a plateaux region indicating the completion of cross-linking. The final storage modulus is about 10 kPa

4. Select the data collection settings appropriate for the rheometer (in the example data shown, measuring points were taken every 2 s for the duration of the measurement; 500 s) (Fig. 14).

5. Load the material on the plate as described in protocol 3.10, and lower the cone to the measuring position (Fig. 14). *See* **Note 9**.

6. Start the measurement.

7. The measurement as described will take a baseline measurement of storage and loss modulus for the chosen time period (Fig. 14). Oscillation rheology is most useful for monitoring the gelation of a material as a function of some cross-linking stimulus (e.g., light) or temperature change.

(a) To perform in situ photorheology, set up the rheometer with a light source routed through an optical fiber, and illuminating the underside of the sample through a quartz crystal stage. It is important to control the UV intensity at the sample using a UV meter. Turn on the UV exposure at some known time point during the measurement—follow the cross-linking via the changing storage and loss moduli of the sample. The "gelation point" is defined as the point at which the storage modulus increases above the loss modulus.

(b) To monitor the gelation of the material as a function of temperature, choose to ramp the temperature (e.g., from 37 °C to 2 °C over the duration of the measurement). Take care to choose a rate of temperature change which is slow enough that the sample tracks the temperature of the measuring system: typically not more than 2 °C/min.

3.4.3 Mechanical Testing: Unconfined Compression

1. Set up the mechanical testing equipment according to the equipment SOP.

2. Mount the chosen load cell on the support stage.

3. Assemble compression plate on top of the load cell. Lift the support stage to near its maximum height and screw it in position.

4. Assemble the top compression plate, using spacers as necessary so that the plate is approx. 5 mm from the bottom plate (Fig. 15).

5. Define a preset value of 3 N relative to the baseline.

6. Place a glass slide on the bottom plate and select the preset force. The top plate will move down and compress the slide until it reaches a force of 3 N. Record the displacement at this preset force. This displacement will represent the base of samples to be compressed. Use the same slide for holding each sample during compression.

7. Return the top mover to a safe distance (e.g., 5 mm gap).

8. In the software, navigate to waveform setup and add/edit the waveform commands, to produce a ramp rate of 0.01 mm/s for a relative movement distance of 2 mm.

9. Area Measurement.

(a) Turn on stereomicroscope. Capture a calibration image using an object of known dimensions (e.g., the scale of a micrometer).

(b) Place hydrogel sample on a slide and place under stereomicroscope. Capture image.

10. Place the glass slide with the test material at the center of the bottom plate. *See* **Note 10**.

11. Drip a few PBS droplets around the sample to maintain its moisture.

12. Adjust the mover so that it is ~1 mm above the sample.

13. Run the measurement. The top plate will slowly move down and compress the sample, while the compression force is recorded.

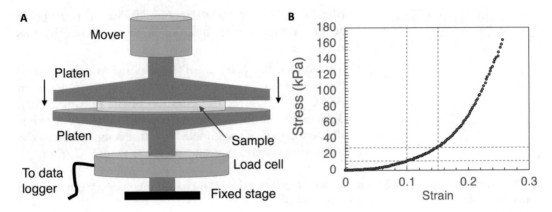

Fig. 15 Left, compressive modulus measurement. Right, typical stress-strain curve for a photocrosslinked hydrogel, compressive modulus 36 kPa

14. Analysis: area measurement. Use image analysis software, such as FIJI/ImageJ (http://imagej.net/Fiji/Downloads) to measure the cross sectional area of the sample from the captured stereomicroscope image. This area measurement is used to convert the "force" data to "stress" data. (Stress = force/area).

15. Analysis: sample thickness. The sample thickness can be obtained from the force-distance curve by noting the position where the force begins to deviate from the baseline (indicating the "top" of the sample) and the known position of the glass slide—the "bottom" of the sample. Subtract one from the other to get the sample thickness. This sample thickness is needed to convert the "displacement" data to "strain" data. (Strain = displacement/sample thickness).

16. Using the values measured in **steps 14** and **15**, convert the measured force-distance curve to a stress-strain curve.

17. The compressive modulus is the slope of the stress–strain curve across some region of interest, for example, the slope between 10 and 15% strain (Fig. 15). *See* **Note 11**.

4   Notes

1. Bioink loading, and handling in general, is easiest when the material is in a liquid state. Many hydrogels exhibit temperature dependent viscosity. Usually, hydrogels are more liquid at higher temperatures, and bioink transfer is best performed at 37 °C. However there are exceptions. Hydrogels based on Pluronic F-127 are more liquid at cold temperatures, and should be handled at a temperature of about 4 °C just after removal from the refrigerator.

2. Tapered nozzles are preferred for bioprinting as the design reduces the shear stresses experienced by encapsulated cells.

3. For many hydrogels, filament formation can be temperature dependent. Thus, ensure the material has reached a stable temperature after the loading process before extruding the filament. The filament test can also be performed at a range of bioink temperatures, if the bioprinter possesses temperature control of the bioink cartridge. Use rheology (Subheading 3.4.2) in a temperature sweep mode to assess the thermal gelation point of hydrogels. Generally, a bioink will only form a filament morphology after gelation.

4. To achieve filament morphology, a thickener additive may be required. Typical thickeners in 3D bioprinting are gelatin and hyaluronic acid. Temperature control can also be used to assist gelation.

5. Many hydrogels exhibit temperature dependent viscosity. When printing at room temperature, care must be taken that the bioink has reached a stable state after removal from storage in the refrigerator, or incubator. Even when the material has reached a stable temperature, further physical cross-linking may take place which alter the viscosity of the material over time—and this will affect printing consistency. For consistent printing, a bioink must exhibit a stable extrusion rate over the timescale of a printing experiment. An alternative strategy is to use motors, rather than air pressure, as the driver for the extrusion [15, 16].

6. The parameter optimization procedure is primarily effective for bioinks whose extrusion rate is stable and reproducible across multiple printing experiments performed on different days. This does not apply to all bioinks (e.g., due to the various sources of variability in 3D bioprinting (*see* **Note 12**). For bioinks that exhibit a relatively variability in extrusion rates greater than about <10%, it is not possible to preoptimize every parameter and hold those parameters across a series of experiments performed over days or weeks. In this case, the filament extrusion rate measurement should be performed before each experiment, and used to adjust the extrusion pressure so that the filament extrusion speed matches that of a preoptimized condition (e.g., 5 mm/s).

7. Programming a 3D bioprinter typically requires knowledge of g-code: a simple programming language based on the Cartesian coordinate system. Many g-code tutorials exist online, and a working knowledge of G-code can typically be gained in a single afternoon, even for researchers with zero programming experience. There are also online G-code toolpath simulators for testing new g-codes (Fig. 16).

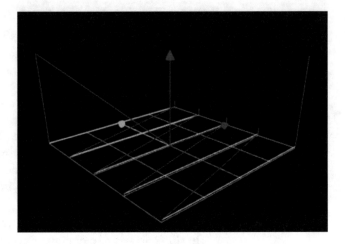

Fig. 16 Example code as visualized using an online g-code simulator

```
Example g-code
; 20 mm lines printed at different z-heights
; Speed: F600 mm/min
; For Nozzle Diameter: 250 um

G21; set units to millimetres
G90; use absolute coordinates
G0 X-10.0 Y10.0 Z10.0; move at maximum speed to position
(x,y,z)=(-10,10,10)
F600; set movement speed

;Line 1
G0 Z0.05; layer height 0.05 mm
M760 ; start extrusion
G1 X10.0 Y10.0; Move at defined speed to position (x,y)=
(10,10)
M761; stop extrusion
G0 Z1.0; raise during movement to next position

;Line 2
G0 Z0.1; layer height 0.1 mm
G0 X-10.0 Y5.0
M760
G1 X10.0 Y5.0
M761
G0 Z1.0

;Line 3
G0 Z0.15; layer height 0.15 mm
G0 X-10.0 Y0.0
M760
G1 X10.0 Y0.0
```

```
M761
G0 Z1.0

;Line 4
G0 Z0.2; layer height 0.2 mm
G0 X-10.0 Y-5.0
M760
G1 X10.0 Y-5.0
M761
G0 Z1.0

;Line 5
G0 Z0.2; layer height 0.2 mm
G0 X-10.0 Y-10.0
M760
G1 X-10.0 Y-10.0
G1 X10.0 Y-10.0
M761


    G0 X0.0 Y0.0 Z10; return to home position 10mm above the
surface
    M84; disable motors
```

8. Consistent loading of the sample materiel is a critical for reproducible results. Minimize the shear forces applied to material
   during loading by using a 1 mL syringe (without a nozzle) to
   control the dispense volume, and using a careful, slow dispensing rate. Between measurements store the material at an appropriate temperature to minimize the formation of physical cross-
   links in the gel (e.g., in a 37 °C incubator for most hydrogels,
   or a 4 °C refrigerator for pluronics).

9. For measurements running longer than ~5 min, drying of the
   hydrogel can introduce artifacts in the measurement. Minimize
   these artifacts by maintaining a humid environment around the
   sample during the measurement, that is, use a rheometer
   equipped with a hood around the measuring system. Droplets
   of water can be dispensed near to the test material to maintain
   high humidity in the local environment. Alternatively, a thin
   film of low viscosity silicone oil can be dispensed around the
   rim of the measuring cone—encapsulating the test material,
   and so preventing evaporation. The silicone oil used must have
   at least ×10 lower viscosity that the test material.

10. The test sample must have very flat surfaces on both the top
    and bottom. Samples for mechanical testing should be cast in
    circular molds with dimensions on the order of 1 cm diameter
    and 2 mm thickness.

11. The region of the stress-strain curve is not standardized across all tissue engineering fields. Researchers targeting some tissues tend to choose other regions (such as that between 0 and 10% strain). Ensure your measurement is comparable to the literature in your field.

12. Some sources of variability in 3D bioprinting:

Source	Details	Minimization
Hardware		
Pressure regulation accuracy	Accuracy of pressure regulation greater than 5% of extrusion pressure can introduce significant variability	Insert another pressure regulator between compressed air supply and the printer
X-Y movement accuracy	Low resolution stepping (e.g., <50 μm) can affect minimum achievable line spacing	Increase line spacing
Z-movement and calibration accuracy	Z-accuracy should be at least 1/10th of the nozzle diameter for reproducible results	Use larger nozzle diameters
Levelness of platform	Can result in prints smeared on one side of build plate, and with poor adhesion on the other	Relevel platform relative to print-head axis
Consumables		
Nozzle variability	Plastic tapered nozzles have variation in their internal diameter, especially nozzles of 0.2 mm inner diameter or smaller	Use plastic nozzles of 0.25 mm inner diameter or greater. Use stainless steel nozzles
Substrate smoothness/ adhesion	Poor ink–substrate compatibility can affect deposition of the first layer	Use surface modification amenable to bioink adhesion
Bioink		
Temperature stability	Many hydrogels exhibit temperature dependent viscosity	Characterize using temperature sweep rheology. Use a motor to drive extrusion rather than air pressure
Temporal stability	Some hydrogels (e.g., gelatin) form physical cross-links over time, gradually increasing viscosity	Characterize using oscillatory rheology

(continued)

Source	Details	Minimization
Homogeneity	Clumpiness of bioink can affect extrusion stability or clog the nozzle altogether	Ensure complete dissolution of all bioink components. Prefilter components
Environmental		
Temperature	Temperature variability in the lab can exacerbate issues with temperature dependent viscosity	Control the temperature of the bioprinting cartridge if possible. Maintain consistent lab temperature
Humidity	Printed hydrogel structures can begin to dry out during a long print run	Humidify local printing environment. Shorten printing times

References

1. Vacanti JP, Langer R (1999) Tissue engineering: the design and fabrication of living replacement devices for surgical reconstruction and transplantation. Lancet 354:S32–S34

2. Kang H-W, Lee SJ, Ko IK et al (2016) A 3D bioprinting system to produce human-scale tissue constructs with structural integrity. Nat Biotechnol 34:312–319

3. Hospodiuk M, Dey M, Sosnoski D et al (2017) The bioink: a comprehensive review on bioprintable materials. Biotechnol Adv 35:217–239

4. Tibbitt MW, Anseth KS (2009) Hydrogels as extracellular matrix mimics for 3D cell culture. Biotechnol Bioeng 103(4):655–663

5. Therriault D, White SR, J a L (2003) Chaotic mixing in three-dimensional microvascular networks fabricated by direct-write assembly. Nat Mater 2:265–271

6. Kolesky DB, Truby RL, Gladman AS et al (2014) 3D bioprinting of vascularized, heterogeneous cell-laden tissue constructs. Adv Mater 26:3124–3130

7. Discher DE, Janmey P, Wang Y (2005) Tissue cells feel and respond to the stiffness of their substrate. Science 310:1139–1143

8. Engler AJ, M a G, Sen S et al (2004) Myotubes differentiate optimally on substrates with tissue-like stiffness: pathological implications for soft or stiff microenvironments. J Cell Biol 166:877–887

9. Saha K, Keung AJ, Irwin EF et al (2008) Substrate modulus directs neural stem cell behavior. Biophys J 95:4426–4438

10. Gilbert F, O'Connell CD, Mladenovska T et al (2017) Print me an organ? Ethical and regulatory issues emerging from 3D bioprinting in medicine. Sci Eng Ethics 24(1):73–91

11. Schuurman W, Levett PA, Pot MW et al (2013) Gelatin-methacrylamide hydrogels as potential biomaterials for fabrication of tissue-engineered cartilage constructs. Macromol Biosci 13:551–561

12. Paxton N, Smolan W, Böck T et al (2017) Proposal to assess printability of bioinks for extrusion-based bioprinting and evaluation of rheological properties governing bioprintability. Biofabrication 9(4):044107

13. O'Connell CD, Di Bella C, Thompson F et al (2016) Development of the Biopen: a handheld device for surgical printing of adipose stem cells at a chondral wound site. Biofabrication 8:15019

14. He Y, Yang F, Zhao H et al (2016) Research on the printability of hydrogels in 3D bioprinting. Sci Rep 6:29977

15. Duchi S, Onofrillo C, O'Connell CD et al (2017) Handheld co-axial bioprinting: application to in situ surgical cartilage repair. Sci Rep 7:5837

16. Di Bella C, Duchi S, O'Connell CD et al (2017) In situ handheld three-dimensional bioprinting for cartilage regeneration. J Tissue Eng Regen Med 12(3):611–621

Chapter 8

Laser-Assisted Bioprinting for Bone Repair

Davit Hakobyan, Olivia Kerouredan, Murielle Remy, Nathalie Dusserre,
Chantal Medina, Raphael Devillard, Jean-Christophe Fricain,
and Hugo Oliveira

Abstract

Bioprinting is a novel technological approach that has the potential to solve unmet questions in the field of tissue engineering. Laser-assisted bioprinting (LAB), due to its unprecedented cell printing resolution and precision, is an attractive tool for the in situ printing of a bone substitute. Here, we describe the protocol for LAB and its use for the in situ bioprinting of mesenchymal stromal cells, associated with collagen and nanohydroxyapatite, in order to favor bone regeneration in a calvaria defect model in mice.

Key words Laser-assisted bioprinting, Bone regeneration, Mesenchymal stromal cells, Regenerative medicine

1 Introduction

LAB technology emerged from the initial work of researchers at the Naval Research Laboratory [1], as a tool to engineer artificial tissues. This printing method is based on the laser-induced forward transfer (LIFT) effect. LIFT-based bioprinters or LAB are composed of three main components: (1) a pulsed laser source, (2) a target, or ribbon, a base structure from which the biological material is printed, and (3) a receiving substrate that collects the printed material. In brief, the target is composed by a laser transparent support (i.e., glass or quartz) of a laser-absorbing material, obtained by plasma treatment of gold or titanium. The organic phase to be printed consists of biomolecules and/or cells suspended on a liquid (e.g., culture media or a hydrogel-like collagen), which is then homogeneously deposited at the surface of the metal film. This organic phase is designated bioink. The laser pulse, controlled by the predefined computer-assisted design (CAD) and by the scanning mirror, will induce the vaporization of the metal film, which results on the production of a droplet, and that is

Jeremy M. Crook (ed.), *3D Bioprinting: Principles and Protocols*, Methods in Molecular Biology, vol. 2140,
https://doi.org/10.1007/978-1-0716-0520-2_8, © Springer Science+Business Media, LLC, part of Springer Nature 2020

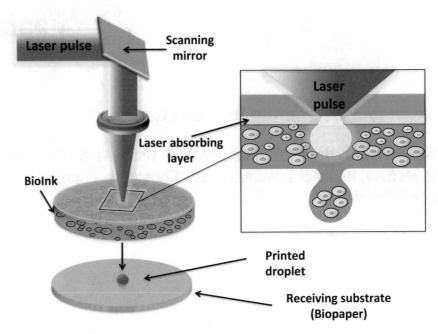

Fig. 1 Schematic representation of the laser-assisted bioprinting (LAB) technology. A standard LAB setup consists of a pulsed laser beam, a focusing system, a ribbon (transparent glass slide) coated with a laser-absorbing layer of metal, onto which a thin layer of bioink is spread, and a receiving substrate facing the ribbon. The physical principle of LAB is based on the generation of a cavitation-like bubble, in the bioink film, whose expansion and collapse induces the formation of a jet and, thereby, the transfer of the bioink from the ribbon to the substrate. (Adapted with permission from [4])

deposited onto the facing substrate (Fig. 1) [2, 3]. LAB by a direct-write method can, at the MHz speed range, deposit droplets containing cells or biomaterials within liquid phase.

Owing to its resolutions, at the picoliter-level, LAB allows to print highly defined cell geometries with 3D organization, enabling unprecedented control over cell behavior and fate, both key parameters in tissue engineering. As such, LAB is an emerging and promising technology to fabricate tissue-like structures with the capacity to mimic the physiological functionality of their native counterparts. Additionally, this method has additional advantages such as automation, reproducibility, and high throughput, making it compatible with the fabrication of 3D constructs of physiologically relevant sizes.

We have recently shown the potential of LAB to attain the in situ printing of mesenchymal stromal cells, associated with collagen and nanohydroxyapatite, in a calvaria bone defect model in mice, and demonstrated to favor bone regeneration. Here, we describe the procedure, step-by-step, of in situ LAB in a calvaria bone defect in mice.

2 Materials

2.1 Reagents

Unless mentioned otherwise, all reagents were obtained from Sigma-Aldrich and were of analytical grade.

2.2 Laser-Assisted Bioprinting Workstation

In general, laser-assisted bioprinting (LAB) setup may be simplified as an assembly of (1) a substrate to store the material of interest, (2) a tool (laser beam) to transfer the material in a desired pattern, and (3) deposition of the material onto a substrate that properly preserves the imprint. Importantly, for proper bioprinting process (including optimal resolution, throughput, reproducibility, automation, and cell viability), each of the abovementioned components should be thoroughly tested. In particular, the LAB workstation used in a previous study [4] was described in detail [5]. In addition, in reference [6] we describe cell-patterning protocols using the workstation. Here, we offer a complimentary and thorough description of the application of the workstation for the in vivo bone repair study reported at [4].

2.2.1 Computer-Assisted Design

1. Workstation integrated software: AST07-LAB (Novalase, France). Defines a printing configuration using "tool" command, which embodies laser beam-related parameters—diaphragm aperture, pattern scanning rate, laser power, and laser pulse rate. In addition, the software allows adding sample patterns like dot, line, circle, rectangle, etc. To finish the printing configuration, add trajectory by assigning pattern-targeting coordinates and define a printing "process" command. Note that several tools may be embedded in a single process command, which will be manifested consecutively.

2. After setting laser beam-related parameters, the tool command may be completed by adding the pattern encrypted in the 8-bit bmp image. Here, the software will numerically scan the image and assign a pattern to the zeros (black) values in the image. Complete printing configuration by adding trajectory and forming a process command.

3. For convoluted patterns, the software allows pattern coordinates assignment using text files, which should be encrypted with the tool command and trajectory to from printing process.

2.2.2 Donor Carrier

1. A motorized rotation stage (carousel) allows selection of the donor substrate (*see* Subheading 2.3) among five positions of donor support plates, enabling the printing of different bioink types (multicolor printing).

2. Donor support plate: A metallic support that holds 30 mm diameter and 1.5 mm thick glass substrate (*see* Subheading 2.3). The support has a greater outer radius to be held by carrousel.

3. At this step *see* Subheading 3.1 for the description of the bioink deposition procedure.

4. Place the substrate on the donor support, ensure that gold-coated side is facing outward of support and that the substrate is properly held by clamps, maintaining its parallel position to the support.

2.2.3 Laser-Induced Forward Transfer

1. Laser source: Nd:YAG crystal laser, Navigator I, Newport Spectra Physics, with 1064 nm wavelength, 30 ns pulse width, 1–100 kHz repetition rate range, 6 W average power.

2. Optical setup: Standard optical mirrors and lenses are used to guide the laser beam toward the scanning system and to set a proper diameter at the entrance of scanning system. The latter is composed of two galvanometric mirrors (SCANgine 14, ScanLab), which can deflect the light beam with up to 2000 mm/s scan rate within microradian angular resolution.

3. Focusing tool: F-theta lens with 58 mm effective focal length (S4LFT, Sill Optics) that enables a constant focal spot within a relatively large image plane.

4. To transfer the pattern, operate at room temperature (RT) while keeping several hundreds of micrometers between the donor and receiver. Tune the laser power (6–80 mW) at 1 kHz repetition rate.

2.2.4 Biopaper Receiver

1. In vitro experiment: A metallic support for the 30 mm diameter glass substrate spread with biopaper (*see* Subheading 2.5).

2. In vivo experiment: A metallic holder designed for mouse experimentation (Fig. 3).

3. Place the receiver on the motorized 3D stage, which has a micrometer range displacement step and 25 mm linear travel range.

2.3 Donor Slide

An essential element of LAB is the substrate that allows proper storage and printability of desired material. Such substrate is commonly referred as a donor. Here, the donor is an optically transparent glass (B270 glass) slide, deposited with a sacrificial gold layer and bioink containing the material of interest (*see* Subheading 2.4). The glass has IR transmittance ≈92.5%, which is required to efficiently transmit light through the substrate and focus on coated material. The choice of metal is justified by its biocompatibility, to avoid cell harm, and absorbance within the optical penetration depth-coated layer at considered laser wavelength, enabling jet formation.

1. Clean glass substrate with detergent and gently wipe with optical tissue.

2. Thoroughly rinse the substrate with ultrapure water.

3. Sterilize the substrate with 70% (v/v) ethanol in ultrapure water, gently removing liquid with optical tissue.

4. Collect the substrate into reservoir and place inside drying sterilizer Poupinel at 80 °C for 4 h.

5. Place the substrate inside a chamber of coater (Emscope SC 500), on top of anode facing the gold cathode for plasma-enhanced sputter deposition. To obtain ≈50 nm thick layer, run the process for 2 min at 6 Pa Argon pressure.

6. Carefully collect the substrate, preserving its sterility and the gold layer. If sterile conditions are not preserved, rinse with 70% (v/v) ethanol and dry in oven for a several min at 80 °C.

2.4 Bioink

We have used the multipotent mouse bone marrow stromal precursor D1 cell line (ATCC) as part of the bioink. Cells are cultured on TCPS plastic suing Dulbecco's Modified Eagle Medium (DMEM, Gibco, Life Technologies), supplemented with 10% (v/v) fetal calf serum (FCS) (Lonza) at 37 °C and with 5% CO_2, using standard cell culture methods.

To monitor cells, upon in vitro and in vivo printing, D1 cells are infected with lentiviral vectors containing the TdTomato protein gene (red fluorescence), under the control of the phosphoglycerate kinase promoter [7]. For viral transduction of this cell type, 2×10^5 freshly trypsinized D1 cells are exposed to 6×10^6 viral particles (multiplicity of infection (MOI) = 30) for 24 h. Cells can then be expanded over several passages, using standard cell culture procedures. Transduction efficiency (>90%) should be evaluated by standard flow cytometry analysis.

1. The medium of subconfluent (approximately 70%) D1 cell cultures is removed and the cell monolayer is washed with phosphate-buffered saline without calcium or magnesium (dPBS, Gibco) and then incubated with Trypsin/EDTA (0.025% trypsin and 0.01% EDTA in PBS, Gibco) at 37 °C for 3–5 min. For a standard 150 cm^2 culture flask, 2 mL of trypsin/EDTA solution is sufficient to efficiently detach cells.

2. Cells are resuspended in complete medium (DMEM with 10% (v/v) FCS). As FCS neutralizes the action of trypsin, adjust the volume of medium according to volume of trypsin used. Typically use 10 mL of complete medium for 2 mL of trypsin. As cell homogeneity is crucial for optimal LAB printing, special care should be taken in order to attain homogeneous suspension of cells at this step.

3. Determine cell concentration and viability by diluting twofold with trypan blue (0.4% w/v solution in PBS, Sigma) and subsequent counting using a standard cell counting chamber (e.g., Neubauer chamber).

4. Centrifuge cells at $300 \times g$ for 5 min, remove the supernatant, and resuspend D1 cells at 120×10^6 cells/mL in DMEM, supplemented with 10% (v/v) of FCS. Special care should be taken to avoid bubble formation. Bioink should be used fresh.

2.5 BioPaper

Type I collagen is used as the receiving substrate, termed the biopaper. The biopaper is prepared as follows:

1. Collagen type I from rat tail (concentration range 6–8 mg/mL, depending on batch; BD Biosciences) is neutralized and diluted to 2 mg/mL using DMEM. Positive-displacement pipettes should be used in this step (Gilson Microman Pipette). While the formation of bubbles should be avoided, they can be removed by brief centrifugation at 4 °C. After neutralization, collagen will gel. At 37 °C, gelation will take approximately 20–30 min. If the solution is not used immediately, it should be kept on ice for up to 1 h after which it will start to gel.

2. Due to its osteoinductive properties [8], nanohydroxyapatite is mixed with the neutralized collagen gel at 1.2% (w/v) and well dispersed using positive-displacement pipettes. For the production of the nanohydroxyapatite using wet chemical precipitation, please refer to published method [9].

3 Methods

The following procedures should be carried at RT, unless otherwise mentioned.

3.1 In Vitro Bioprinting

1. For in vitro printing, deposit and homogeneously spread 100 μL of a neutralized and cold solution of type I rat collagen solution (2 mg/mL), as described in Subheading 2.5, point 1, over a 3-cm diameter round glass slide (previously sterilized). Prior to use, the glass slide should be placed in a 37 °C incubator for 2 h in order to allow the collagen to gel. This procedure is performed in a laminar flow hood, following standard sterile operation procedures.

2. Deposit 30 μL of the bioink solution (prepared as described in Subheading 2.4), and using the exterior of the pipette tip, spread evenly over the donor slide (in contact with the gold layer). Special care should be taken not to scratch the gold layer.

3. Place both the donor slide and the receiving substrate in the LAB setup and execute the desired protocol in terms of geometry, distance between the receiver and donor slides, and energy and frequency of the laser pulses.

day 0   **day 2**   **day 4**

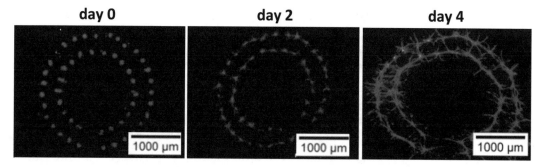

Fig. 2 Representative fluorescence images of ring printed tomato-positive D1 cells at days 0, 2, and 4. (Adapted with permission from [4])

4. After bioprinting, place the receiving slide (biopaper) in a 6-well plate, in a humidified incubator with 5% CO_2, in order to allow cellular attachment. This step is highly dependent on cellular type. D1 cells attach to the collagen substrate in 20–30 min. Evaluate the optimal attachment time for different cell types (cells should change morphology and grow into the collagen layer).

5. Slowly add cell culture medium to the well and culture using standard conditions. From this step, several evaluations can be performed to evaluate geometry conformity, cellular viability, proliferation, and organization over time. *See* Fig. 2 for an example of D1 cell printing pattern (two circle geometry) and subsequent follow-up by microscopy up to 4 days.

3.2   In Vivo Bioprinting

All procedures for animal handling should follow the principles of laboratory animal care established by responsible entities applicable to the country or region where the experiments are performed. Additionally, experiments should be carried out in accredited animal facilities following international, national, or regional regulations (e.g., European recommendations for laboratory animal care, directive 86/609 CEE of 24/11/86).

1. Prior to performing surgical procedure, prepare bioinks (*see* Subheading 2.4) and the biopaper (collagen, *see* Subheading 2.5). In the bioprinting room, keep the bioinks under a cell culture hood at RT and maintain the collagen solution on ice.

2. In the surgerical room, install surgical equipment on a sterile field (classic surgical tray: compressors, antiseptic agent, 0.9% sodium chloride solution, eye drop solution, surgical tape, scalpel, dissecting forceps, trephine, micromotor, scissors, absorbable suture).

3. Prepare anesthetic syringes for intraperitoneal injection of mice with a mix of Ketamin (Ketamine 1000, Virbac, France), Xylazine (Rompun, Bayer, France), and Buprenorphine

(Buprécare, Axience, France). Only adult mice are used in such in vivo experiments. The working solution (1 mL) is based on average body weight (e.g., 25 g: 11.5 mg Ketamine, 1.15 mg Xylazine, 0.010 mg Buprenorphine).

4. Anesthetize the mouse with an appropriate anesthetic dose, based on weight: 90 mg/kg for Ketamine, 9 mg/kg for Xylazine, 0.08 mg/kg for Buprenorphine.

5. Administer eye drop to avoid ocular complications. Renew the application of eye drop solution as necessary until the end of the procedure.

6. Remove the hair from the surgical site with electric hair clippers and apply depilatory cream for 2 min before cleaning and rinsing.

7. Use povidone iodine (Betadine) for skin antisepsis onto the surgical area.

8. Make an incision in the middle of the skull, from the nasal bone to the superior nuchal line, in order to expose the calvaria.

9. Carefully peel off the periosteum membrane using the scalpel.

10. Perform two lateral calvaria bone defects with a bone trephine of 3 mm.

11. Irrigate the surgical site constantly with a solution of sodium chloride 0.9% (w/v) during surgical procedure.

12. Dry the print area with sterile compresses.

13. Perform a simple suture on each side of the head to keep the wound open.

14. Transport the mouse to the laser-bioprinting room.

15. Allow the collagen to gel for 5 min.

16. Protect the eyes of the mouse from the laser beam with surgical tape.

17. Place the mouse on the stereotaxic rack (specific mouse holder). Use surgical tape to immobilize the mouse and fold down its ears to avoid any contact with the ribbon during the printing process. Only one defect is used for laser printing procedure while the contralateral site is used as control. Dura mater must be as horizontal as possible in order to precisely face the ribbon.

18. Introduce the mouse on its holder inside the bioprinting workstation onto (x,y,z) motorized translation stages.

19. Focus the CCD camera on the dura mater by translating mouse holder according to the z-axis. This position is then recorded as the origin of z-position.

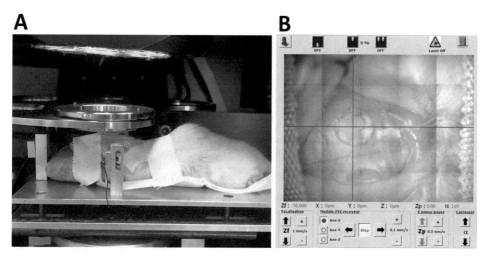

Fig. 3 (**a**) LAB setup for in vivo bioprinting. (**b**) LAB operating system photo, depicting the calibration of the bioprinting positioning in relation to the mice calvaria bone defect

20. Target the center of the defect through the touchscreen. This position is then recorded as the origin of printed pattern (Fig. 3a, b for a snapshot of both the LAB setup and of the operation software interface, enabling the positioning of the mice and the calibration of the printing spot).

21. Select the desired pattern, laser parameters (energy and frequency), and printing gap distance in the software.

22. Spread manually the bioink on the donor slide using a pipette and place it in the workstation on the dedicated rack.

23. Close the door of the laser workstation to insure laser safety.

24. Start printing.

25. Print a layer of neutralized collagen containing 1.2% nanohydroxyapatite (biopaper) on a disk design of 2 mm, with 50 impact spots, repeated three times, in order to attain a disk of approximately 100 μm.

26. Print the gel geometry, for example, a ring based on 50 impact spots, where each impact contained around 50 cells is shown (Fig. 2). Typically, using the described LAB printer, printing settings used are 300 μm/s speed, 1 kHz frequency, laser energy of 27.5 μJ, and a gap distance (between the ribbon and the receiver layer) of 1000 μm.

27. Reproduce the pattern as many times as necessary.

28. Once printed, check the accuracy of the printing pattern under high magnification (binocular loupe). The use of fluorescent cells is recommended to allow visualization of the pattern.

29. Again, print a layer of neutralized collagen containing 1.2% nano hydroxyapatite (biopaper) on a disk design of 2 mm, with 50 impact spots, repeated three times.

30. Allow the collagen to jellify for 5 min.

31. Bring the mouse back to the surgical room.

32. Reposition soft tissues and suture the wound.

33. During postsurgical recovery, place the mouse on a heated water blanket and monitor the animal constantly until awake.

34. After recovery, return the mouse back to the animal facility.

Acknowledgment

The authors acknowledge the financial support given by Inserm (France) in the framework of the "Accelerateur de Recherche Technologique" ART BioPrint, the "Fondation des Gueules Cassées" (N°54-2017), and the "Fondation de l'Avenir," Paris, France (N°AP-RM-17-038).

References

1. Barron JA, Wu P, Ladouceur HD, Ringeisen BR (2004) Biological laser printing: a novel technique for creating heterogeneous 3-dimensional cell patterns. Biomed Microdevices 6(2):139–147

2. Guillemot F, Souquet A, Catros S, Guillotin B (2010) Laser-assisted cell printing: principle, physical parameters versus cell fate and perspectives in tissue engineering. Nanomedicine 5 (3):507–515. https://doi.org/10.2217/nnm. 10.14

3. Ringeisen BR, Othon CM, Barron JA, Young D, Spargo BJ (2006) Jet-based methods to print living cells. Biotechnol J 1(9):930–948. https://doi.org/10.1002/biot.200600058

4. Keriquel V, Oliveira H, Remy M, Ziane S, Delmond S, Rousseau B, Rey S, Catros S, Amedee J, Guillemot F, Fricain JC (2017) In situ printing of mesenchymal stromal cells, by laser-assisted bioprinting, for in vivo bone regeneration applications. Sci Rep 7(1):1778. https://doi.org/10.1038/s41598-017-01914-x

5. Guillemot F, Souquet A, Catros S, Guillotin B, Lopez J, Faucon M, Pippenger B, Bareille R, Remy M, Bellance S, Chabassier P, Fricain JC, Amedee J (2010) High-throughput laser printing of cells and biomaterials for tissue engineering. Acta Biomater 6(7):2494–2500. https://doi.org/10.1016/j.actbio.2009.09.029

6. Devillard R, Pages E, Correa MM, Keriquel V, Remy M, Kalisky J, Ali M, Guillotin B, Guillemot F (2014) Cell patterning by laser-assisted bioprinting. Methods Cell Biol 119:159–174. https://doi.org/10.1016/B978-0-12-416742-1.00009-3

7. Shaner NC, Campbell RE, Steinbach PA, Giepmans BN, Palmer AE, Tsien RY (2004) Improved monomeric red, orange and yellow fluorescent proteins derived from Discosoma sp. red fluorescent protein. Nat Biotechnol 22 (12):1567–1572. https://doi.org/10.1038/nbt1037

8. Fricain JC, Schlaubitz S, Le Visage C, Arnault I, Derkaoui SM, Siadous R, Catros S, Lalande C, Bareille R, Renard M, Fabre T, Cornet S, Durand M, Leonard A, Sahraoui N, Letourneur D, Amedee J (2013) A nano-hydroxyapatite—pullulan/dextran polysaccharide composite macroporous material for bone tissue engineering. Biomaterials 34 (12):2947–2959. https://doi.org/10.1016/j. biomaterials.2013.01.049

9. Keriquel V, Guillemot F, Arnault I, Guillotin B, Miraux S, Amedee J, Fricain JC, Catros S (2010) In vivo bioprinting for computer- and robotic-assisted medical intervention: preliminary study in mice. Biofabrication 2(1):014101. https://doi.org/10.1088/1758-5082/2/1/014101

Chapter 9

Bioprinting Stem Cells in Hydrogel for In Situ Surgical Application: A Case for Articular Cartilage

Serena Duchi, Carmine Onofrillo, Cathal O'Connell, Gordon G. Wallace, Peter Choong, and Claudia Di Bella

Abstract

Three-dimensional (3D) bioprinting is driving major innovations in the area of cartilage tissue engineering. As an alternative to computer-aided 3D printing, in situ additive manufacturing has the advantage of matching the geometry of the defect to be repaired without specific preliminary image analysis, shaping the bioscaffold within the defect, and achieving the best possible contact between the bioscaffold and the host tissue. Here, we describe an in situ approach that allows 3D bioprinting of human adipose-derived stem cells (hADSCs) laden in 10%GelMa/2%HAMa (GelMa/HAMa) hydrogel. We use coaxial extrusion to obtain a core/shell bioscaffold with high cell viability, as well as adequate mechanical properties for articular cartilage regeneration and repair.

Key words Cartilage regeneration, Coaxial 3D extrusion, Core-shell geometry, GelMa/HAMa hydrogel, Human adipose-derived stem cells, In situ photocrosslinking

1 Introduction

3D bioprinting is driving major innovations in the area of cartilage tissue engineering [1]. A key challenge is to bring cumbersome bench-based technology to the operating room for real-time application. 3D printing of scaffolds for chondral and osteochondral repair has been tried, using a mosaicplasty-type technique, based on the implantation of scaffolds in defects that are premade in the laboratory before surgery [2]. Despite using the most advanced preoperative imaging techniques, however, a mismatch between the bioscaffold and the host defect will always remain because (1) surgical debridement before implantation alters the anatomy previously imaged and (2) small gaps within damaged cartilage fail to be identified.

The graft-site mismatch and the size-depth mismatch between implanted scaffolds and surrounding tissues are the driving causes

Jeremy M. Crook (ed.), *3D Bioprinting: Principles and Protocols*, Methods in Molecular Biology, vol. 2140,
https://doi.org/10.1007/978-1-0716-0520-2_9, © Springer Science+Business Media, LLC, part of Springer Nature 2020

for the deterioration of implanted cartilage, which in turn leads to implant failure [3].

As an alternative to the computer-aided 3D printing, in situ additive manufacturing or in situ 3D bioprinting has been proposed as a valid option to deliver at the time and point of need a bioscaffold capable of regenerating the deficient tissue and to limit the graft-site mismatch [4, 5]. The advantages of in situ biomanufactoring for cartilage regeneration include the possibility of perfectly matching the defect geometry without specific preliminary image analysis, shaping the bioscaffold within the defect and achieving the best possible contact between the bioscaffold and the host tissue. What we propose as an in situ approach is an advanced handheld coaxial extrusion device that allows the deposition of cells embedded in a hydrogel material in the surgical setting [6]. Through this device, stem cell-laden hydrogel scaffolds can be extruded in a coaxial core/shell distribution, which favorably affects cell survival while maintaining adequate mechanical properties for cartilage tissue engineering purposes. This result is achieved, thanks to the segregation of the cells in the core away from the photocrosslinking reaction confined to the shell compartment [7]. Photocrosslinking is one of the most used strategies to obtain a stiff hydrogel. To bring this technique to surgical application and make it compatible with surgical timings, photocrosslinking cannot last longer than a few seconds. Therefore, a relatively high voltage is required. In our system, the core component of the coaxial hydrogel contains infrapatellar hADSCs encapsulated within a naturally derived hydrogel (GelMa/HAMa, 10%/2%). The outer shell component of this coaxial system contains the same hydrogel (GelMa/HAMa, 10%/2%), which becomes photopolymerizable due to the addition of lithium-acylphosphinate (LAP) as photoinitiator (PI). Hardening of the shell provides the structural properties that allow 3D printing, while the hADSCs are preserved in a relatively soft, cell-friendly environment inside the core. This system allows the generation of core/shell GelMa/HAMa bioscaffolds with stiffness of 200 KPa, achieved after only 10 s of exposure to 700 mW/cm^2 of 365 nm UV-A, containing >90% viable stem cells that retain proliferative capacity.

Overall, the core/shell handheld 3D bioprinting strategy enabled rapid generation of high modulus bioscaffolds with high cell viability, with potential for in situ surgical cartilage engineering.

2 Materials

2.1 Isolation and Cultivation of hADSCs

1. Dulbecco's phosphate-buffered saline (PBS) 1×.

2. Red cell lysis buffer, Sigma-Aldrich (St. Louis, MO, USA).

3. 0.1% EDTA/0.25% trypsin, Sigma-Aldrich (St. Louis, MO, USA).

4. Collagenase type II 345 U/m, Worthington Biochemical Corporation (Lakewood, NJ, USA) (*see* **Note 1**).

5. Fetal bovine serum (FBS).

6. Antibiotic (penicillin/streptomycin).

7. L-Glutammine 200 mM.

8. Hepes 1 M.

9. Human epidermal growth factor (hEGF), R&D Systems Inc. (Minneapolis, MN, USA) 10 μg/ml.

10. Human fibroblastic growth factor-2 (hFGF), R&D Systems Inc. (Minneapolis, MN, USA) 10 μg/ml.

11. Cell culture medium: Dulbecco's modified Eagle's medium 1000 mg/ml glucose, #D6046, Sigma-Aldrich (St. Louis, MO, USA).

12. Complete culture medium: DMEM low glucose + 1% antibiotics + 2.5% Hepes (15 mM) + 1% glutamime (2 mM) + 10% FBS + 20 ng/ml hEGF + 1 ng/ml hFGF.

13. Disposable scalpels.

14. Nylon mesh, 100 and 40 μm, Millipore (Darmstadt, Germany).

15. Tissue culture plastic: sterile pipets, culture flasks.

16. Glass petri dish.

17. CO_2 incubator.

18. Centrifuge.

19. Inverted light microscope.

20. Hemocytometer/CellCountessII ThermoFisher.

2.2 GelMa and HAMa Materials

1. Gelatin from porcine skin, gel strength ~300 g bloom, type A (Sigma-Aldrich).

2. Hyaluronic acid (HA, 1200–1900 kDa).

3. Dulbecco's PBS 1× and Pen/Strep.

4. MilliQ water.

5. Methacrylic anhydride.

6. Schott bottles.

7. Sterile 50 ml centrifuge tubes with vented caps, 0.2 μm pore size (corning 50 ml mini bioreactor, Sigma-Aldrich, cat. no. CLS431720).

8. Sterile syringes (50 ml volume).

9. Syringe filters (0.22 μm).

10. Cell strainer (70 or 40 μm).

11. 1000- or 500-ml Schott bottle.

12. Sterile 5-ml yellow-capped container.

13. Freeze-dryer.

14. Pipettor and tips.

15. Electronic balance.

16. Aluminum foil.

17. Thermal Benchtop Incubated Shaker.

2.3 Coaxial Bioprinting

1. 3D-printed titanium coaxial nozzle.

2. Dulbecco's PBS 1× and Pen/Strep.

3. Electronic balance.

4. Tweezers ×2, sterile.

5. Glass coverslips, sterile.

6. 1- and 5-ml syringes, sterile.

7. 3-ml pipettes, sterile.

8. Cartridge ×2, sterile.

9. Blue cap cartridge ×2, sterile.

10. White plunger cartridge ×, sterile.

11. Glass petri dish, sterile.

12. 12-Well plate.

13. Yellow-capped containers 5 ml × 2.

14. Magnetic stirring bar, sterile.

15. Temperature controlled oven.

16. Complete hADSCs cell culture media.

17. PDMS molds (diameter 7 mm, height 2 mm), cleaned with ethanol 70%.

18. Spatula, sterile.

19. GelMa/HAMa 10%/2% (core).

20. GelMa/HAMa 10%/2% + LAP 0.1% (shell).

21. Cells 10×10^6/ml hADSCs.

22. Photoinitiator: lithium-acylphosphinate (LAP) from Tokyo Chemical Industry Co. (Tokyo, Japan). Use 0.1% in GelMa/ HAMa hydrogel.

23. Light irradiation source: 36 5 nm UV Omnicure LX400+, Lumen Dynamics LDGI, fitted with a 12-mm lens (25-mm focal distance) at maximum intensity.

24. Dual air pressure controller.

25. Air pump.

26. Water bath sonicator.

3 Methods

All steps are performed at room temperature (RT) in sterile condition under a laminar flow hood.

3.1 hADSCs Isolation and Culture

1. Place the infrapatellar fat pad (IPFP) excised during surgery using a scalpel in a sterile container containing PBS1× + PenStrep1× (*see* **Note 2**).

2. Under a laminar flow hood, place the IPFP in a sterile glass petri dish and add 3 ml of complete culture medium. Using forceps and scalpel, remove the fat from the fibrous material and dice the fat until only the fibers are left.

3. Collect the minced fat material with a plastic sterile pipette and place in a clean 50-ml falcon tube (*see* **Note 3**).

4. Add 0.3 ml of 10 mg/ml collagenase II solution (final concentration is 1 mg/ml) for 3 ml of media (adjust volume if more media has been used on **step 3**), and place for 3 h at 37 °C on rotating platform at 160 rpm.

5. Centrifuge 2100 × g for 10 min.

6. Discard the supernatant, resuspend the pellet in 5 ml of PBS1× + Pen/Strep, and filter through a mesh nylon filter 100 μm.

7. Centrifuge 400 × g (400 rcf) for 5 min.

8. Discard the supernatant and resuspend the pellet in 5 ml of Red Cell Lysis buffer for 10 min at RT.

9. Filter through a mesh nylon filter 40 μm.

10. Centrifuge 400 × g (400 rcf) for 5 min.

11. Discard the supernatant and resuspend the pellet in 0.5 ml of complete culture medium.

12. Count cells using hemocytometer counting.

13. Plate the cells in a T25 flask.

14. After 48 h, the first cell colonies should appear attached on the bottom of the flask. Wash cells vigorously in PBS1×, discard and add new complete culture media.

15. After 7 days, cells should be all 80–90% confluent. So detach them with trypsin for 3 min at 37 °C, centrifuge at 300 × g for 3 min, plate them at 4000–5000 cell/cm^2 density for expansion, and freeze several vials for future experiments/characterization.

3.2 Preparation of GelMa

1. Dissolve 10 g in 100 ml sterile PBS1× at 50 °C and stir until fully dissolved (3 h).

2. Add 8 ml methacrylic anhydride gradually while the pH is maintained at 4.4 by dosing of alkaline solution using a pH-controlled dosing pump.

3. Allow the reaction to proceed for 3 h at 50 °C.

4. Dilute with 300 ml PBS1× to terminate the methacrylation reaction.

5. Dialyze the solution in distilled water (MWCO 12–14 kDa) at 40 °C for 7 days.

6. After purification, freeze dry the GelMa solution (clear colorless viscous liquid). It should result in a bright white product with 65%–70% yield.

7. Calculate the degree of functionalization (73%) by 1H-NMR in D_2O [8].

8. Store the freeze-dried product at 4 °C in a dark and inert environment prior to use.

3.3 Sterilization of GelMa: Filtration Method

1. With Schott bottle on scale, add dry GelMa up to a maximum of 15 g and record the mass (see **Note 4**).

2. Add MilliQ water to make a final concentration of 1.5% w/w GelMa (see **Note 5**).

3. Place at 37 °C for 2–3 h. Meanwhile, autoclave a clean 1000 ml Schott bottle.

4. In a class II biological safety cabinet, pour GelMa solution through a 100-μm cell strainer and a 70-μm cell strainer into a 1000-ml Schott bottle.

5. Filter-sterilize the GelMa solution into vented-capped clean 50 ml flacon tubes using 0.2-μm syringe filter units (see **Note 6**).

6. Measure and record the mass (with the highest possible accuracy) of each vented-capped 50 ml centrifuge tube, while it is still empty. Add ~40 ml GelMa solution to centrifuge tube.

7. Place in −80 °C freezer overnight.

8. Transfer all aliquots to the freeze-dryer without allowing the solutions to thaw (see **Note 7**).

9. Lyophilize (freeze-dry) until the GelMa is fully dehydrated, typically 4–7 days (see **Note 8**).

10. Measure and record mass of all dry centrifuge tubes. Subtract the mass of the original (empty) tube to calculate mass of the dry hydrogel inside. Label each tube with the mass of its contents.

11. In a biosafety cabinet, exchange vented caps with standard screw-top caps to avoid hygroscopic absorption of water during storage (see **Note 9**).

12. Store material at −80 °C until needed.

3.4 Preparation of HAMa

1. Dissolve 0.5 g of HA overnight in sterile deionized water (120 ml).

2. Add 3.75 ml methacrylic anhydride (5 M equivalent of the hydroxyl groups per HA disaccharide repeating unit).

3. Allow the reaction to proceed by stirring at RT overnight with the pH maintained at 8.0 by adding 5 N NaOH (3.75 ml, 5 M equivalent of the hydroxyl groups per HA disaccharide repeating unit).

4. Dialyze the HAMa (MWCO12–14 kDa) for 48 h.

5. After purification, freeze-dry the GelMa solution (clear color-less viscous liquid).

6. Calculate the degree of functionalization (~20%) by 1H-NMRinD2O [8].

7. Store the freeze-dried product at 4 °C in a dark and inert environment, prior to use.

3.5 Sterilization of HAMa: Chloroform Method

1. With Schott bottle on scale, add dry HAMa up to a maximum of 5 g and record the mass (*see* **Note 10**).

2. Add MilliQ water to make a final concentration of 0.5% w/w HAMa.

3. Place in incubator-shaker at 37 °C, shaking at ~70 rpm. Leave until completely dissolved (i.e., solution is clear), typically ~2–3 days.

4. Carefully layer chloroform at the bottom of the hydrogel solu-tion. The amount of chloroform to use should be ~10% of the volume of hydrogel solution. DO NOT SHAKE OR STIR. Allow to stand overnight at 2–8 °C.

5. Next morning, the hydrogel has gelled overnight. To remove the chloroform, use a sterile serological pipette to pierce through the gel. Remove as much as possible from the bottom of the Schott bottle and place in chloroform waste. Some immiscible drops may remain, but they will be removed in the next steps.

6. Measure and record the mass (with the highest possible accu-racy) of each empty, vented-capped 50 ml centrifuge tube, while it is still empty.

7. Add ~40 ml of hydrogel solution to centrifuge tube.

8. Place in −80 °C freezer overnight.

9. Transfer all aliquots to the freeze-dryer without allowing the solutions to thaw (*see* **Note 7**).

10. Lyophilize (freeze-dry) until the HAMA is fully dehydrated, typically 4–7 days. You can repeatedly measure the mass of one of the samples to monitor the freeze-drying process (*see* **Note 8**).

11. Measure and record mass of all dry centrifuge tubes. Subtract the mass of the original (empty) tube to calculate mass of the dry hydrogel inside. Label each tube with the mass of its contents.

12. In a biosafety cabinet, exchange vented caps with standard screw-top caps to avoid hygroscopic absorption of water during storage.

13. Store material at $-80\ ^{\circ}$C until needed.

3.6 Preparation of GelMa/HAMa

1. Take the freeze-dried GelMa and HAMa aliquots required from $-80\ ^{\circ}$C freezer.

2. Clean the electronic scale, place an empty 50-ml yellow-capped container on the balance and set the weight to be 0 (tare).

3. Under a biosafety hood, transfer GelMa and HAMa into the sterile 50-ml yellow-capped container, add a sterile magnetic stirring bar and weigh it again on the scale. Record the weight and subtract the tare.

4. Calculate the amount of PBS1×-Pen/Strep solution required according to concentration desired: amount of PBS1×-Pen/Strep required = mass of GelMa + HAMa/concentration (weight %). For GelMa/HAMA 10%/2%: 1 g GelMa + 0.2 g HAMa in 10 ml of PBS1×-Pen/Strep.

5. Add calculated amount of PBS1×-Pen/Strep into the container with pipettor.

6. Slightly shake the container, label it with date, chemical name, concentration, and name of the operator and wrap it with aluminum foil.

7. Keep the wrapped container in the Thermal Benchtop Incubated Shaker at 37 °C–70 rpm until the solution is completely dissolved, typically 3–4 days. It should result in a clear yellowish viscous liquid with no clumps.

8. Store the hydrogel in the fridge (4 °C) for further use (leave the stirring bar in the container).

3.7 Coaxial Bioprinting

1. The day before printing, remove GelMa/HAMa from Subheading 3.5 from the fridge and keep in the oven at 37 °C, stir overnight. In the meanwhile, immerse the PDMS molds in ethanol (70%) overnight (*see* **Note 11**).

2. The day of the printing: activate the pressure pump and set up the controller to generate around 30 kPa for the extrusion of the GelMa/HAMa in the core and 15 kPa for the extrusion of the GelMa/HAMa in the shell.

3. Turn on the UV light source by setting the conditions of time, power, and temperature (Fig. 1a) (*see* **Note 12**).

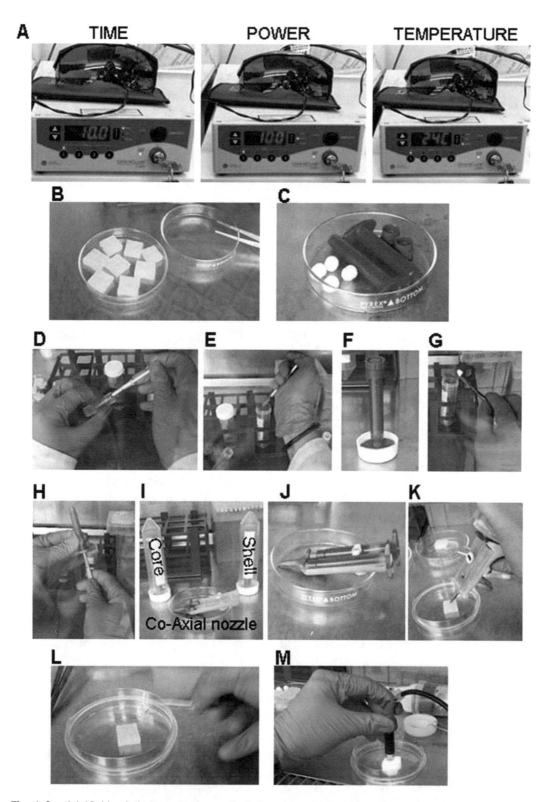

Fig. 1 Coaxial 3D bioprinting preparation and printing steps. (**a**) Light irradiation UV source: the indicated conditions (time, power, temperature) are designed to generate 700 mW/cm². (**b**) PDMS molds. (**c**) Cartridges, blue caps, and white plungers. (**d–h**) Steps showing the loading of the GelMa/HAMa hydrogel into core and shell cartridges. (**i, j**) Loading of the core and shell cartridges in the coaxial nozzle. (**k–m**) Coaxial bioprinting steps: bioprinting in molds (**k**), covering samples with a sterile glass coverslip (**l**), and irradiating with the UV light (**m**)

4. Remove the PDMS molds from ethanol and leave to dry on glass dish (Fig. 1b).

5. Prepare cartridges, blue caps and white plungers (Fig. 1c).

6. Take the GelMa/HAMa hydrogel from the incubator, and under a biosafety hood, divide in two sterile 5 ml yellow-capped containers.

7. Add LAP to GelMa/HAMa (shell) to obtain 100 µl/ml GelMa-HAMa (0.1% LAP), and resuspend using 1 ml syringe, avoiding bubbles (Fig. 1d).

8. Transfer the shell onto the cartridge with the 1 ml syringe (Fig. 1e).

9. Close the tip of the cartridge with a blue cap, and turn it upside down, so the hydrogel will gently sleep toward the bottom (Fig. 1f).

10. Place the white plunger inside the cartridge, but do NOT PUSH THROUGH (Fig. 1g) and leave the cartridge in the oven at 37 °C while preparing the core.

11. Detach and count the cells and calculate how many cells per ml of material are required, according to the specific application or type of experiment.

12. Add the cells to GelMa/HAMa to obtain 10×10^6/ml, and resuspend carefully using 1 ml syringe, avoiding bubbles (core) (Fig. 1d).

13. Close the tip of the cartridge with a blue cap, and turn it upside down, so the hydrogel will gently sleep toward the bottom (Fig. 1f).

14. Place the white plunger inside the cartridge, but do NOT PUSH THROUGH (Fig. 1g).

15. Take the shell from the incubator, and prepare to mount the two cartridges on the coaxial nozzle. Remove the blue cap and then push the white plunger with the help of a tweezer until the biomaterial fills the tip of the cartridges (Fig. 1h).

16. Load core and shell cartridges to short screw of the coaxial nozzle (Fig. 1i) (see **Note 13**).

17. Once the cartridges are properly screwed (Fig. 1j), connect to the pressure pump system and test the extrusion until the hydrogel is extruded from both compartments of the nozzle.

18. Print samples in molds (Fig. 1k), cover samples with a sterile glass coverslip (Fig. 1l), and irradiate to photocrosslink with the UV light for 10 s (Fig. 1m) (see **Note 14**).

19. Remove sample from mold with the help of spatula, remove the glass coverslip, move the samples in in a sterile 12-well plate, and add 1 ml of PBS1×-Pen/Strep sterile.

20. Remove the PBS1× and add 1 ml of complete hADSCs growth media, and transfer the plate in a CO_2 incubator (*see* **Note 15**).

3.8 Cleaning of the Coaxial Nozzle Before and After Use

1. Submerge the nozzle in a beaker containing 10% bleach for at least 30 min (*see* **Note 16**).

2. Clean it vigorously by passing 1–5 ml of 10% bleach through the core and shell compartments, one by one, with a syringe, until the fluid comes out from both sides.

3. Wash vigorously with distilled water to remove the bleach with 1 ml syringe through both compartments.

4. Sonicate for 30 min by immersing the nozzle in a water bath sonicator, but do not let the metal get in contact with it, use amplitude of 65%, and keep the pulse on.

5. After sonication, perform a standard test of the flow rate with the pressure pump. Add 3 ml of water to the cartridge in the core, and close with the piston, turn pressure on, and block the shell hole; set 20 kPa, shot air continuously until water runs out, and record the time. Repeat the same for the shell cartridge.
   TEST OK: core = extrusion of 3 ml of water at 20 kPa pressure in less than 8 s; shell = extrusion of 3 ml of water at 20 kPa pressure in less than 4 s.

6. If test passes, sterilize the nozzle in autoclave (121 °C–20 min) in paper folder.

7. Dry it in the oven at 65 °C for 30 min (*see* **Note 17**).

4 Notes

1. Enzyme solution is prepared at 10 mg/ml concentration in PBS1×, filtered through a sterile 0.22 μm filter, aliquoted and stored at −20 °C.

2. Preferentially use within 1 h from the surgery or store at 4 °C overnight. Add cell culture media or PBS1× containing antibiotics if no synovial fluid is present in the collecting container from the surgery room.

3. Add more cell culture media if necessary to clean up and collect all the minced fat from the petri dish.

4. This document describes a protocol for sterilizing GelMa or HAMa beginning with dry material. It could potentially be modified to be performed just after synthesis (before initial freeze-drying step). It is based on the methods recommended by Loessner et al. [9] and by Sigma-Aldrich: http://www.sigmaaldrich.com/content/dam/sigmaaldrich/docs/Sigma/General_Information/1/c5533inf.pdf.

5. MilliQ at RT has a density of 0.998 g/ml, so we can assume this equates to 1.5% w/v to high precision.

6. Hand-filtering results in loss of $13.2 \pm 1.22\%$ of GelMa when a 1.5% solution is filtered).

7. Transfer from freezer to freeze-dryer in dry ice if necessary.

8. You can repeatedly measure the mass of one of the samples to monitor the freeze-drying process. In our experience, we measured a mass-loss rate of approx. 0.6 g/h for my samples. When repeated mass measurements (several hours apart) give the same value, the process is complete.

9. The vented caps can be reused with new 50 ml falcon tubes.

10. HAMa is too viscous, even at very low concentrations to filter, so the chloroform method is preferable compared to the filtration.

11. To cast PDMS mold, you can 3D print the negative with PLA and prepare shape and size of your preference according to the specific application.

12. The indicated conditions are designed to generate 700 mW/cm$_2$, so timer is 10 s, power 100, and temperature 24 °C (Fig. 1a).

13. After closing the system, check the extrusion from the nozzle, and any possible leakage by pushing carefully the white plunger with a sterile tweezer.

14. Make sure the temperature of the UV light does not go above 30 °C, otherwise wait until it cools down to repeat the irradiation on the other samples.

15. The sample's volume for 2 mm thick, 7 mm diameter is around 77 μl/sample, and the cell number/sample is 4×10^5 cells.

16. Commercial bleach from supermarket is fine.

17. If the test is not passed, clean and sonicate again.

Acknowledgment

The authors would like to acknowledge the following funding sources: Foundation for Surgery John Loewenthal, Royal Australasian College of Surgeons, Research Endowment Funds, St Vincent's Hospital Melbourne, Arthritis Australia, Eventide Homes Research Project Grant, and The CASS Foundation Research Grant.

References

1. Daly AC, Freeman FE, Gonzalez-Fernandez T, Critchley SE, Nulty J, Kelly DJ (2017) 3D bioprinting for cartilage and osteochondral tissue engineering. Adv Healthc Mater 6:1–20

2. Rai V, Dilisio MF, Dietz NE, Agrawal DK (2017) Recent strategies in cartilage repair: A systemic review of the scaffold development and tissue engineering. J Biomed Mater Res A 105:2343–2354

3. Mollon B, Kandel R, Chahal J, Theodoropoulos J (2013) The clinical status of cartilage tissue regeneration in humans. Osteoarthritis Cartilage 21:1824–1833

4. Ozbolat IT (2018) Bioprinting scale-up tissue and organ constructs for transplantation. Trends Biotechnol 33:395–400

5. Keriquel V, Guillemot F, Arnault I, Guillotin B, Miraux S, Amédée J et al (2010) In vivo bioprinting for computer- and robotic-assisted medical intervention: preliminary study in mice. Biofabrication 2:14101

6. O'Connell CD, Di Bella C, Thompson F, Augustine C, Beirne S, Cornock R et al (2016) Development of the Biopen: a handheld device for surgical printing of adipose stem cells at a chondral wound site. Biofabrication 8:15019

7. Duchi S, Onofrillo C, O'Connell CD, Blanchard R, Augustine C, Quigley AF et al (2017) Handheld co-axial bioprinting: application to in situ surgical cartilage repair. Sci Rep 7:1–12

8. Burdick JA, Chung C, Jia X, Randolph MA, Langer R (2005) Controlled degradation and mechanical behavior of photopolymerized hyaluronic acid networks. Biomacromolecules 6:386–391

9. Loessner D, Meinert C, Kaemmerer E, Martine LC, Yue K, Levett PA et al (2016) Functionalization, preparation and use of cell-laden gelatin methacryloyl-based hydrogels as modular tissue culture platforms. Nat Protoc 11:727–746

Chapter 10

3D Bioprinting Electrically Conductive Bioink with Human Neural Stem Cells for Human Neural Tissues

Eva Tomaskovic-Crook and Jeremy M. Crook

Abstract

Bioprinting cells with an electrically conductive bioink provides an opportunity to produce three-dimensional (3D) cell-laden constructs with the option of electrically stimulating cells in situ during and after tissue development. We and others have demonstrated the use of electrical stimulation (ES) to influence cell behavior and function for a more biomimetic approach to tissue engineering. Here, we detail a previously published method for 3D printing an electrically conductive bioink with human neural stem cells (hNSCs) that are subsequently differentiated. The differentiated tissue constructs comprise functional neurons and supporting neuroglia and are amenable to ES for the purposeful modulation of neural activity. Importantly, the method could be adapted to fabricate and stimulate neural and nonneural tissues from other cell types, with the potential to be applied for both research- and clinical-product development.

Key words 3D bioprinting, Electrically conductive bioink, Stem cells, Human neural tissue, Electrical stimulation

1 Introduction

Until recently, a lack of physiologically relevant preclinical models of the human central nervous system (CNS) has hampered our understanding and ability to treat associated diseases and injury. This is underscored by a decline in pharmaceutical research and development efficiency, resulting in a decrease in new drug launches in the last several decades [1]. To begin with, the selection of targets for novel drug therapies necessitates a comprehensive understanding of human disease biology, from the etiological factors, ensuing pathological mechanisms, and clinical effects. An important corollary of bona fide drug targets is the discovery of new diagnostic biomarkers to classify patient populations with homogenous etiologies and pathophysiologies, circumventing the poor specificity and sensitivity of traditional approaches. Neoteric human biological systems such as brain organoids, neuroids, and other multicellular and multilayered neural culture systems are

Jeremy M. Crook (ed.), *3D Bioprinting: Principles and Protocols*, Methods in Molecular Biology, vol. 2140,
https://doi.org/10.1007/978-1-0716-0520-2_10, © Springer Science+Business Media, LLC, part of Springer Nature 2020

incontrovertibly advancing insight by better recapitulating CNS form and function, inclusive of remarkably complex functional structure, toward superior studies of pathologies and drugs, and increased therapeutic reach. This is particularly important to studying complicated neurological diseases where there is a growing body of evidence for them being unique to humans [2].

Bioprinting for human neural tissue engineering offers exciting opportunities in both research and translation, including modeling human tissue development, function, and dysfunction, as well as responsivity to pharmaceuticals such as neuroleptics, and replacement tissues for diseased, aged, or injured CNS. The potential to create more physiologically relevant live-tissues with the hallmark cellular and molecular componentry of native CNS is gaining interest world wide, with printer and associated technologies rapidly advancing.

Here, we describe in detail a protocol adapted from previous research publications on human neural tissue engineering using human neural stem cells (hNSCs), a novel electrically conductive biogel, and 3D bioprinting [3, 4]. Our presently explained approach incorporates commercially available midbrain-derived hNSCs combined with the biogel to form a printable bioink that is chemically cross-linked for a mechanically robust 3D cell-laden scaffold construct. Importantly, further details of the bioink properties including mechanical, physical, and electrical properties (*see* **Note 1**) can be found in our earlier publications [3, 4]. Moreover, the protocol describes the methods for in situ hNSC proliferation, differentiation, and characterization of resulting neural tissues. Finally, although the protocol depicts the fabrication of neural tissue from hNSCs, it could conceivably be adapted to create other tissues using different stem cells.

2   Materials

2.1 Bioink Preparation

1. Alginic acid sodium salt from brown algae (alginate, Al; Sigma-Aldrich) (*see* **Note 2**).
2. Carboxymethyl chitosan (CMC; Santa Cruz Biotechnology).
3. Agarose (Ag; low gelling temperature; Sigma-Aldrich).
4. Phosphate-buffered saline (PBS), pH 7.4 (Sigma-Aldrich).
5. Autoclaved 20 mL glass vial and magnetic stir bar.
6. 19 G needle.
7. 5 mL syringe.

2.2 hNSC Culture and Differentiation

1. hNSCs (e.g., ReNcell VM human neural progenitor cell line; Millipore).

2. hNSC proliferation medium: Complete NeuroCult Proliferation Medium comprising NeuroCult NS-A Basal Medium (human; STEMCELL Technologies) and NeuroCult NS-A Proliferation Supplement (human; STEMCELL Technologies), supplemented with heparin (2 μg/mL; Sigma-Aldrich), epidermal growth factor (EGF, 20 ng/mL; Peprotech), and basic fibroblast growth factor (bFGF, 20 ng/mL; Peprotech) (*see* **Note 3**).

3. hNSC differentiation medium: Dulbecco's Modified Eagle Medium (DMEM)/Ham's F-12 Nutrient Mixture (F-12): Neurobasal Medium, 1:1 (v/v) (Gibco, Life Technologies) supplemented with 1% NeuroCult SM1 neuronal supplement (STEMCELL Technologies), 0.5% N2 Supplement-A (STEMCELL Technologies), 1× GlutaMAX Supplement (Gibco, Life Technologies), and brain-derived neurotrophic factor (BDNF; 50 ng/mL; Peprotech),

4. Laminin (Gibco, Life Technologies).

5. TrypLE Select (Gibco, Life Technologies).

6. Trypan Blue (Sigma-Aldrich).

7. PBS, pH 7.4 (Sigma-Aldrich).

8. Standard 6-well plates (Corning).

9. 15 mL conical centrifuge tubes (Corning).

2.3 Bioprinting

1. 3D computer-assisted design software, e.g., Blender open source software.

2. Smoothflow tapered dispensing tips, e.g., 27 G, 0.2 mm internal diameter (Nordson EFD).

3. 30 cc optimum syringe barrel and piston with tip-end and snap-on end caps (Nordson, EFD).

4. Standard 12-well culture plates (Corning).

5. Calcium chloride (Sigma-Aldrich).

6. PBS, pH 7.4 (Sigma-Aldrich).

7. Dulbecco's Modified Eagle Medium (DMEM)/Ham's F-12 Nutrient Mixture (F-12).

2.4 hNSC Viability Assay

1. Calcein AM (Life Technologies).

2. Propidium iodide (PI, Life Technologies).

3. Tissue culture dish with coverslip bottom (170 μm thickness).

4. Image J (Fiji) software.

2.5 Membrane Permeant Live Cell Neural Labeling

1. NeuroFluor NeuO (STEMCELL Technologies).

2. hNSC differentiation medium (*see* Subheading 2.2, **item 3**).

3. Tissue culture dish with coverslip bottom (170 μm thickness).

2.6 Immuno-phenotyping

1. 3.7% (w/v) paraformaldehyde (PFA; Sigma-Aldrich) solution in PBS (pH 7.4).

2. Goat serum (Sigma-Aldrich).

3. Triton X-100 (Sigma-Aldrich).

4. PBS, pH 7.4 (Sigma-Aldrich).

5. Unconjugated primary antibodies:

    (a) Nestin (NES; mouse; Millipore).

    (b) SRY (sex-determining region Y)-box 2 (SOX2; rabbit; Millipore).

    (c) Vimentin (VIM; mouse; Millipore).

    (d) Marker of proliferation ki67 (MKI67; rabbit; Abcam).

    (e) Neuron-specific class III β-tubulin (TUJ1; mouse; Covance).

    (f) Glial fibrillary acidic protein (GFAP; rabbit; Millipore).

    (g) Microtubule-associated protein 2 (MAP2; mouse, Millipore).

    (h) Synaptophysin (SYP; rabbit; Abcam).

6. Secondary antibodies:

    (a) Goat antimouse IgG (H+L) highly cross-adsorbed secondary antibody, Alexa Fluor 488 (Life Technologies).

    (b) Goat antirabbit IgG (H+L) highly cross-adsorbed secondary antibody, Alexa Fluor 594 (Life Technologies).

7. 4, 6-diamidino-2-phenylindole (DAPI; Life Technologies).

8. Prolong Gold antifade reagent (Life Technologies).

9. Tissue culture dish with coverslip bottom (170 μm thickness) or glass slides fitted with silicone spacer and coverslip (170 μm thickness).

10. Image J (Fiji) software.

2.7 Live Cell Calcium Imaging

1. Fluo-4 AM (Life Technologies).

2. HEPES buffered Hank's balanced salt solution (HHBS; 137 mM NaCl, 5.4 mM KCl, 1.5 mM $CaCl_2$, 0.4 mM $MgSO_4$, 0.5 mM $MgCl_2$, 0.3 mM Na_2HPO_4, 0.44 mM KH_2PO_4, 4 mM $NaHCO_3$, 5.6 mM D-glucose, and 10 mM HEPES (4-(2-hydroxyethyl)-1-piperazineethanesulfonic acid, pH 7.4).

3. Bicuculline (Sigma-Aldrich).

4. Tissue culture dish with coverslip bottom (170 μm thickness).

5. Image J (Fiji) software.

2.8 Equipment	1. Magnetic stirrer with heating.

2. CO_2 incubator.

3. 3D-Bioplotter® System (EnvisionTEC GmbH) equipped with platform and dispensing head temperature control and contained within sterile biological safety cabinet.

4. Laser Scanning Confocal (Leica TCS SP5 II) equipped with an inverted microscope and Leica Application Suite for Advanced Fluorescence (LAS AF) software.

3 Methods

Preparation of reagents and cell culture should be performed in a sterile biosafety cabinet. Unless otherwise stated, media aliquots and reagents should be prewarmed to 37 °C. Incubations and culturing should be performed in a 37 °C incubator with a humidified atmosphere of 5% CO_2 in air.

3.1 Bioink Preparation

1. Dissolve 1.5% (w/v) Ag in PBS in a 20-mL glass vial by heating in a microwave oven for 5 s. Allow it to cool and allow bubbles to disperse. Repeat five to eight times for 1–2 s each time until fully dissolved (*see* **Note 4**).

2. Add 5% (w/v) CMC and stir on a magnetic stirrer with hotplate set at 60 °C for 5 min or until transparent.

3. Add 5% (w/v) Al and continue stirring with a magnetic stirrer at 60 °C for 1 h or until dissolved (*see* **Note 5**).

4. Allow the final solution to cool to 37 °C in dry incubator and store at temperature until introduction of cells (*see* **Note 6**).

3.2 hNSC Monolayer Cell Culture and Passaging

Although hNSC can be maintained as neurosphere suspension cultures, we routinely expand hNSC as adherent monolayer cultures as outlined below. For further information regarding culturing hNSC as neurospheres, refer to the protocol described elsewhere [5].

1. Prepare hNSC proliferation medium as described in Subheading 2.2.

2. Prepare laminin-coated 6-well plates the day prior to hNSC seeding. Thaw laminin aliquot at 4 °C to prevent gelation. Dilute laminin to 10 μg/mL in cold PBS, pH 7.4. Add 1 mL of 10 μg/mL laminin to each well of a 6-well plate. Wrap up the plate with Parafilm to prevent evaporation and incubate overnight at 4 °C. Following overnight incubation, wash wells with 2 mL PBS and store wrapped in Parafilm and foil at 4 °C for up to 2 weeks. Just before harvesting cells for subculture, replace PBS with 2 mL per well of hNSC proliferation medium and warm plate at 37 °C.

3. Adherent monolayer hNSC cultures should be passaged when 60–80% confluent. Discard the supernatant, and add 0.5 mL TrypLE to each well and swirl to distribute. Incubate at 37 °C in a 5% CO_2 humidified incubator for 3 min.

4. Add 1 mL hNSC proliferation media (without ADDS; *see* **Note 3**) to each well to stop the digestion. Gently triturate the cells to achieve single cell suspension (*see* **Note 7**).

5. Centrifuge cell suspension at $190 \times g$ for 3 min, aspirate the medium, and resuspend the cells with fresh hNSC proliferation medium (without ADDS).

6. Determine the number of viable cells using, for example, Trypan Blue exclusion assay and count cells with a hemocytometer.

7. If cells are to be further subcultured or expanded for bioprinting, seed hNSC at a density of 96,000 cells per well of 6-well plate. Return cells to 37 °C in a 5% CO_2 humidified incubator. Perform half media change every 2–3 days.

3.3 Preparation of Cell-Laden Bioink for Bioprinting

Once bioink has been warmed to 37 °C and cells have been freshly harvested and counted to determine the number of viable cells, the cells are mixed into the bioink in preparation for bioprinting:

1. Calculate the total number of viable cells remaining after completion of Subheading 3.2 to determine the total volume of cell-laden bioink to achieve a desired density of 10×10^6 hNSCs per mL of bioink.

2. Using a 5-mL syringe, carefully remove the warm bioink from the glass vial. Dispense the total volume of bioink required as calculated in **step 1** into a sterile tip-end capped 30 cc syringe barrel.

3. Centrifuge hNSCs (harvested in Subheading 3.2) at $190 \times g$ for 3 min to pellet cells and remove supernatant. Gently resuspend pellet in a 10% volume (total volume of cell-laden bioink as calculated in **step 1**) using hNSC proliferation medium (without ADDS). Transfer the total resuspended cell pellet to the sterile tip-end capped syringe barrel containing bioink from **step 2**.

4. Carefully mix the hNSCs into the bioink by stirring with a 19 G needle to obtain an evenly distributed cell-laden bioink. Insert a piston into the syringe barrel and apply a snap-on endcap to seal the syringe.

5. Centrifuge the sealed syringe containing cell-laden bioink at $300 \times g$ for 1 min to remove air bubbles.

6. Cool the syringe containing cell-laden bioink at 4 °C for 10 min prior to printing.

3.4 Bioprinting

The design of the 3D model should be carried out before the day of experiment. Turn on and set the EnvisionTEC 3D Bioplotter instrument to precool (platform and dispensing head) to 15 °C before harvesting of cells. Ensure that the Bioplotter work-space is sterilized by UV exposure and wipe down surfaces with 70% ethanol before use to maintain sterility within biological safety cabinet.

1. Using a computer-assisted design (CAD) software (e.g., Blender or SolidWorks), create a 3D model to bioprint cell-laden biomaterial in a lattice structure (e.g., 5 mm × 5 mm × 2 mm). Generate a stereolithography (STL) file.

2. Using 3D-Bioplotter RP software installed on the Bioplotter, open the saved STL file. Create uniform slice thickness of layers equivalent to 80% of the inner nozzle tip diameter and convert to a BPL format file.

3. Open the BPL file in VisualMachines software to establish material parameters and pattern design for 3D printing.

4. Fit syringe containing cell-laden bioink onto a low temp viscous dispensing head. Remove end-tip cap from syringe, and replace with tapered dispensing tip. Follow the manufacturer's instructions to control the 3D-Bioplotter (*see* **Note 8**).

5. Printing parameters are determined by the end user. As a guide, modify the print speed and dispensing pressure of continuous lines at 90° angle between layers to generate scaffold constructs, having equally sized strands and intervening pores.

6. Once optimal parameters have been determined, extrusion print in air the cell-laden bioink into separate wells of a sterile 12-well culture plate.

7. Immediately following printing of each construct in a well, cover the printed constructs with 2% (w/v) calcium chloride in PBS, being careful not to disrupt the structure of the printed construct, and cross-link bioink for 10 min at room temperature (RT).

8. After cross-linking, rinse constructs three times with 1 min washes in DMEM/F-12.

9. Continue washes with two successive 10 min incubations in hNSC proliferation medium (without ADDS) at 37 °C in a 5% CO_2 humidified incubator.

10. Finally incubate constructs for 1 h in hNSC proliferation medium (without ADDS), followed by a medium change to hNSC proliferation medium and return to incubator for extended tissue culture.

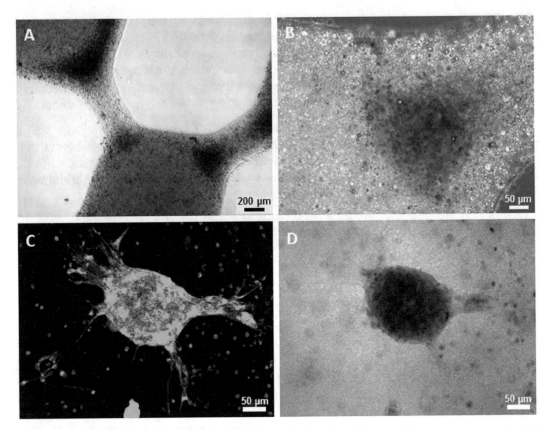

Fig. 1 hNSCs within 3D printed constructs (**a**, **b**; bright-field light microscopy images), and following live cell neural labeling with NeuroFluor NeuO (**c**, **d**; confocal microscopy image and corresponding transmitted light image, respectively)

3.5 Culture and Differentiation of Printed Constructs

1. Incubate constructs in hNSC proliferation medium for a minimum of 4 days post printing, with half-volume medium changes performed every 2–3 days (Fig. 1a, b).

2. Differentiate hNSCs to functional neurons and supporting neuroglia within constructs by replacing proliferation medium with hNSC differentiation medium. Culture for a minimum of 3 days. If continuing differentiation for longer than 3 days, include half-volume medium changes every 3–4 days (*see* **Note 1**). The end user, depending on individual experimental requirements, should determine length of culture of printed constructs for proliferation and differentiation (e.g., 10 days proliferation and 14 days differentiation) [3].

3.6 Assessment of hNSC Viability

1. Assess hNSC viability for days 1 and 7 after printing by incubating printed constructs with 5 μg/mL calcein AM in DMEM/F-12 at 37 °C for 10 min, followed by addition of PI to give a final dilution of 1 μg/mL PI and incubate for 10 min.

2. Remove labeling solution and wash constructs for 10 min in DMEM/F-12.

3. Transfer the constructs into tissue culture dish with coverslip bottom for confocal microscopy.

4. Calculate the number of live and dead cells using ImageJ.

3.7 Membrane Permeant Live Cell Neural Labeling

1. Immediately before use, dilute NeuroFluor NeuO (100 µM stock concentration) in hNSC differentiation medium at 1:800 to achieve working concentration of 0.125 µM (*see* **Note 9**).

2. Aspirate culture medium from constructs and add 1 mL of labeling medium to each well. Protect from light.

3. Incubate at 37 °C in a 5% CO_2 humidified incubator for 1 h.

4. Remove labeling medium. Add 1 mL of fresh hNSC differentiation medium and incubate at 37 °C for a further 2 h.

5. Transfer constructs to a tissue culture dish with coverslip bottom Add 1 mL of fresh hNSC differentiation medium to cover the construct for imaging.

6. Visualize NeuroFluor NeuO labeling using a fluorescent microscope with appropriate filter sets (NeuO Ex/Em: 468/557 nm) (Fig. 1c, d; *see* **Note 10**).

3.8 Immuno-phenotyping

Undifferentiated constructs can be immunophenotyped after a period of hNSC expansion within printed constructs and compared to constructs after a period of differentiation.

1. Aspirate cell medium and rinse constructs with PBS at RT.

2. Fix samples with 3.7% PFA at RT for 30 min and wash twice in PBS.

3. Perform concurrent blocking and permeabilization overnight at 4 °C with 5% (v/v) goat serum in PBS containing 0.3% (v/v) Triton X-100.

4. Prepare antibodies at appropriate concentration (according to manufacturer's guidelines and as confirmed by the end user) in 5% goat serum in PBS. Incubate constructs with unconjugated primary antibodies overnight at 4 °C.

5. Following overnight incubation, wash constructs three times in 0.1% Triton X-100 in PBS.

6. Add species specific Alexa Fluor-conjugated secondary antibody to constructs and incubate in the dark (wrapped in foil) for 2 h at RT.

7. Wash constructs two times in 0.1% Triton X-100 in PBS at RT.

8. Stain with 10 µg/mL DAPI at RT for 10 min.

9. Transfer the constructs into a tissue culture dish with coverslip bottom for immediate imaging with confocal microscopy. Alternatively, mount constructs onto glass slides fitted with a silicone spacer and coverslip applied with Prolong Gold anti-fade reagent. Image with a confocal microscope.

3.9 Live Cell Calcium Imaging

1. Reconstitute Fluo-4 AM in DMSO as per manufacturer's instructions. Prepare Fluo-4 AM at 1-2 µM in hNSC differentiation medium just before use.

2. Remove spent culture medium from well and wash constructs with HHBS. Load cell-laden constructs with freshly prepared 2 µM Fluo-4 AM for 30–60 min at 37 °C (*see* **Note 11**).

3. Following incubation, rinse the constructs with HHBS and allow cells to stabilize for 30 min in the dark before imaging.

4. Mount constructs on coverslip-bottom dishes to view cells by confocal microscopy. Immerse constructs in HHBS (*see* **Note 12**).

5. Using the confocal microscope, excite cells with 494 nm laser and collect images in the range of the dye emission maximum, 506 nm. Collect data under basal conditions to measure spontaneous intracellular calcium release.

6. Examine changes in calcium flux in response to addition of neurotransmitter of interest (e.g., acetylcholine, glutamate). For instance, add 50 µM bicuculline to induce intracellular calcium release and immediately visualize the signal by confocal microscopy (*see* **Note 13**).

7. Collect time course of calcium response (spontaneous and bicuculline induced) for each region (cell) of interest by measuring average fluorescence intensity (normalized to baseline) using LAS AF software.

4   Notes

1. Optional electrical stimulation (ES) can be performed during and/or after differentiation of printed constructs using a method of choice. We have previously used penetrating conducting polymer (CP) microelectrode arrays for 3D ES of hNSC-laden constructs for neural tissue induction [4]. While the setup and stimulation regimen are described by Tomaskovic-Crook et al. [4], briefly, we employed a two-electrode setup including a platinum mesh auxiliary electrode and the CP electrodes penetrating the printed construct acting as working electrodes. We applied balanced biphasic stimulation for 3 days during hNSC differentiation, preceded by initial 4 days proliferation.

2. Alginic acid sodium salt from brown algae are colloidal, poly-uronic acid structural molecules composed of linked α-L-guluronic (G) and β-D-mannuronic (M) residues of varying sequences, depending on seasonality and location from where the brown algae is isolated. Sodium alginates have a wide range of viscosities, M/G residue ratio, and molecular weight chain distribution that affect subsequent gelation properties. While we have previously used low viscosity alginate with 100–300 cP, M/G ratio 1.67, and a molecular weight ~50,000 Da for optimal hNSC survival during extrusion printing, an optimal % (w/v) ratio of sodium alginate within the bioink must be determined.

3. Store prepared hNSC proliferation medium at 4 °C for up to 1 month, before discarding. To minimize unnecessary wastage of growth factors, hNSC proliferation medium without the addition of heparin, EGF, and bFGF (without ADDS), can be made exclusively for the use of washing cells during subculturing.

4. Keep glass vial covered with lid but ajar to ensure release of gasses during microwave heating. Ensure that agarose solution does not bubble over at any point.

5. The viscous bioink can be additionally stirred with a 19 G needle and vortexed to help disperse any clumps of undissolved material. As needed, centrifuge vial at $300 \times g$ for 1 min to clear remnants of viscous solution from the sides of the vial.

6. The bioink can be stored at 4 °C for up to 3 weeks before discarding.

7. Do not flux the cells by pipetting more than 4–5 times.

8. Before printing constructs, it is necessary to calibrate the dispensing tip position, transfer height and needle offset, and calibrate the needle (dispensing tip) according to the manufacturer's instructions. Needle calibration is required with start-up of machine and after each needle change if required during printing.

9. The suggested working concentration of NeuroFluor NeuO is 0.125–0.25 μM. The end user is recommended to titrate the concentration for each application.

10. NeuroFluor NeuO is nonpermeant and cells remain viable following incubation. Cells should be visualized on the same day as labeling, as the signal will diminish over time.

11. Fluo-4 AM cell loading concentration and incubation time should be determined by the end user to obtain fluorescence with optimal signal to noise ratio.

12. HHBS will assist with maintaining cell viability during imaging.

13. Bicuculline is a competitive GABA(A) receptor antagonist that blocks the inhibitory effect of the neurotransmitter GABA and thus isolates the activity of glutamate receptor mediated excitatory postsynaptic potentials.

Acknowledgments

The authors wish to acknowledge funding from the Australian Research Council (ARC) Centre of Excellence Scheme (CE140100012).

References

1. Crook JM, Wallace G, Tomaskovic-Crook E (2015) The potential of induced pluripotent stem cells in models of neurological disorders: implications on future therapy. Expert Rev Neurother 15(3):295–304

2. Dean B (2009) Is schizophrenia the price of human central nervous system complexity? Aust N Z J Psychiatry 43:13–24

3. Gu Q, Tomaskovic-Crook E, Lozano R, Chen Y, Kapsa RM, Zhou Q et al (2016) Functional 3D neural mini-tissues from printed gel-based human neural stem cells. Adv Healthc Mater 5:1429–1438

4. Tomaskovic-Crook E, Zhang P, Ahtiainen A, Kaisvuo H, Lee CY, Beirne S et al (2019) Human neural tissues from neural stem cells using conductive biogel and printed polymer microelectrode arrays for 3D electrical stimulation. Adv Healthc Mater 8(15):1900425

5. Crook JM, Tomaskovic-Crook E (2017) Culturing and cryobanking human neural stem cells. In: Crook JM, Ludwig TE (eds) Stem cell banking: concepts and protocols, Methods in molecular biology, vol 1590. Human Press, New York

Chapter 11

3D Coaxial Bioprinting of Vasculature

Yang Wu, Yahui Zhang, Yin Yu, and Ibrahim T. Ozbolat

Abstract

Development of a suitable vascular network for an efficient mass exchange is crucial to generate three-dimensional (3D) viable and functional thick construct in tissue engineering. Different technologies have been reported for the fabrication of vasculature conduits, such as decellularized tissues and biomaterial-based blood vessels. Recently, bioprinting has also been considered as a promising method in vascular tissue engineering. In this work, human umbilical vein smooth muscle cells (HUVSMCs) were encapsulated in sodium alginate and printed in the form of vasculature conduits using a coaxial nozzle deposition system. Protocols for cell encapsulation and 3D bioprinting are presented. Investigations including dehydration, swelling, degradation characteristics, and patency, permeability, and mechanical properties were also performed and presented to the reader. In addition, in vitro studies such as cell viability and evaluation of extra cellular matrix deposition were performed.

Key words Vascular, Tissue engineering, 3D bioprinting, Human umbilical vein smooth muscle cells, Sodium alginate

1 Introduction

In the last decade, there has been a great advancement in engineering artificial tissues and organs such as skin [1, 2] tendon [3, 4], liver [5], and kidney [6]. However, obstacles have also hindered the development of thick and complex tissues due to the absence of an efficient blood vessel systems for mass exchange. Fabrication of vascular tissues, with biomimicry, unique mechanical properties, and hierarchical organization, is still a critical challenge in tissue engineering [7]. The major biological function of blood vessels is to deliver oxygen and nutrients into tissues and remove waste products away from tissues. They are also able to withstand pressures and shear stresses, regulate blood flow, and, with certain level of permeability, play critical roles in immunological responses [8]. Diameters of blood vessels vary depending on their physiological locations (e.g., microvessels <1 mm, small vessels 1–6 mm, and large vessels >6 mm) [9].

Jeremy M. Crook (ed.), *3D Bioprinting: Principles and Protocols*, Methods in Molecular Biology, vol. 2140,
https://doi.org/10.1007/978-1-0716-0520-2_11, © Springer Science+Business Media, LLC, part of Springer Nature 2020

Vascular tissue engineering is expected to translate the up-to-date knowledge of vascular biology to develop new clinical therapies. The autogenous vascular grafts have major contributions to reconstructive arterial surgeries, which are however limited by inadequate availability [10]. Furthermore, extended time, cost, and additional morbidity are also concerns due to their harvest procedure [11, 12]. Nonbiodegradable polymers, such as polytetrafluoroethylene, polyethylene terephthalate, and polyurethane, have been utilized for fabrication of artificial blood vessles [13]. However, owing to thrombus formation and compliance mismatch, none of these materials have proved suitable for generating grafts less than 6 mm in diameter [14].

Alternatively, various scaffolds have been developed for vasculature regeneration, including decellularized tissues [7, 15], cell sheet conduits [16, 17], biodegradable synthetic polymer-based constructs [18, 19], and natural biomaterial-based blood vessel constructs [20, 21]. In recent years, three-dimensional (3D) bioprinting has advanced as a promising approach for tissue construction, which provides high precision, automation, and flexibility and is capable of involving one or multiple cell types during the fabrication [22–24]. In this chapter, a protocol for coaxial bioprinting of perfusable vasculature conduits with controlled process parameters is presented. The proposed bioprinting system allows to create vasculature conduits of any length within a short fabrication time [25]. In addition, the presented system does not need postfabrication procedure and enables direct bioprinting of complex media exchange networks [26]. This system can be further extended to various applications, including fabrication of scale-up complex tissue models for drug screening and cancer studies [27].

2   Materials

Prepare all solutions using ultrapure deionized (DI) water and analytical grade reagents.

2.1   Sodium Alginate Preparation

1. Sodium alginate solution (Sigma-Aldrich, USA): powder treated with ultraviolet (UV) light for sterilization with three 30 min cycles, 3%, 4%, and 5% (w/v) in sterilized DI water, stir overnight in a magnetic stirrer at room temperature (*see* **Note 1**).

2. $CaCl_2$ solution (Sigma-Aldrich, USA): powder treated with UV light for sterilization with three 30 min cycles, 4% w/v in sterilized DI water, and filtered through 0.2 μm syringe filter.

2.2   Cell Preparation

1. Primary human umbilical vein smooth muscle cells (HUVSMCs) from Invitrogen™ Life Technologies, USA.

2. Cell culture medium: Medium 231 (Invitrogen Life Technologies, USA) supplemented with smooth muscle cell growth supplement (SMGS) (Invitrogen Life Technologies, USA), 10 U/μl penicillin, 10 μg/ml streptomycin, and 2.5 μg/μl fungizone.

3. 0.25% trypsin-EDTA solution.

2.3 Bioprinting Equipment

1. Pneumatic air dispenser (EFD® Nordson, USA).

2. Syringe pump (New Era Pump System Inc., USA).

3. Coaxial needle with a 22-G inner tube (0.71 and 0.41 mm for outer and inner diameters, respectively), a 14-G outer tube (2.11 and 1.69 mm for outer and inner diameters, respectively), and a feed tube (Fig. 1a, b) (*see* **Note 2**).

2.4 Cell Viability Assay

1. Calcein acetoxymethylester (calcein AM, Sigma-Aldrich, USA) and ethidiumhomodimer-2 solution: 1.0 mM each in phosphate buffered saline (PBS).

2.5 Cell Proliferation Assay

1. MTT   [(3-(4,5-dimethylthiazol-2-yl)-2,5-diphenyltetrazolium-bromide)] assay (Invitrogen Life Technologies, USA): 1% (v/v, 5 mg/ml) in PBS.

2.6 Scanning Electron Microscopy (SEM) Imaging

1. Osmium tetroxide: 1% in PBS.

2. Ethanol with different concentrations: 25%, 50%, 75%, 95%, and 100%, v/v in DI water.

2.7 Tissue Histology for Verhoeff-Van Gieson Staining

1. Paraformaldehyde: 4% in PBS.

2. 5% alcoholic hematoxylin: 5 g hematoxylin in 100 ml 100% alcohol (*see* **Note 3**).

3. 10% aqueous ferric chloride: 10 g ferric chloride in 100 ml DI water.

4. Weigert's iodine solution: 2 g potassium iodide and 1 g iodine in 100 ml DI water (*see* **Note 4**).

5. Verhoeff's working solution: 20 ml 5% alcoholic hematoxylin, 8 ml 10% ferric chloride, and 8 ml Weigert's iodine solution (*see* **Note 5**).

6. 2% aqueous ferric chloride: 10 ml 10% ferric chloride from above in 50 ml DI water.

7. 5% aqueous sodium thiosulfate.

8. Van Gieson's counterstain: 5 ml 1% aqueous acid fuchsin in 100 ml saturated aqueous picric acid.

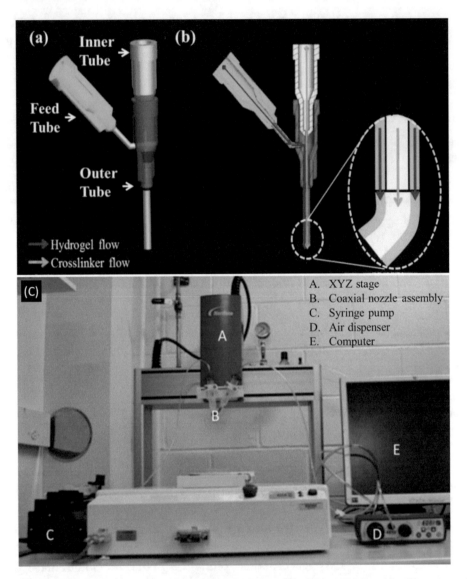

Fig. 1 Experimental setup for vasculature bioprinting: (**a**) 3D model of the coaxial nozzle, (**b**) cross-sectional view of coaxial nozzle assembly model with fluid flow paths for alginate and crosslinker solutions, and (**c**) bioprinter setup for 3D coaxial bioprinting of vasculature

3   Methods

3.1   Cell Preparation

1. Incubate HUVSMCs at 37 °C in 98% humidity and 5% CO_2, change medium every 2 days.

2. Upon reaching 70% confluence, detach cells using a 0.25% trypsin-EDTA solution.

3. Centrifuge cell suspension at $250 \times g$ for 5 min to obtain cell pellets.

4. Resuspend cells in 4% sodium alginate solution to obtain a density of 10×10^6 cells/ml.

5. Gently mix the cell suspension by a vortex mixer to get uniform distribution (*see* **Note 6**).

3.2 Fabrication of Vasculature Conduits

1. Feed the HUVSMCs-loaded alginate to the outer nozzle, and $CaCl_2$ solutions to the inner nozzle.

2. Print the HUVSMCs-loaded alginate at the air pressure of 21 kPa using an air dispenser, and the $CaCl_2$ at a feed rate of 16 ml/min using a syringe pump (Fig. 1c) (*see* **Note 7**).

3. Collect the precrosslinked HUVSMCs-laden alginate conduits in a $CaCl_2$ pool (*see* **Note 8**).

4. Image the conduits using a light microscope (Motic®, BA310, USA) (Fig. 2a).

5. For fabrication of branched conduits, open a window at the wall of the conduit (larger) with a microsurgery scissor, and insert the branch conduit (smaller) (optional).

6. Upon insertion and alignment, seal the branching site with a thin layer of alginate solution containing HUVSCMs (optional).

3.3 SEM Imaging

1. Soak the conduits in a 4% $CaCl_2$ solution for 12 h.

2. Cut the conduits to short sections perpendicular to their longitudinal axis.

3. Rinse the conduits for 10 min in PBS.

4. Fix the cells in conduits using 4% paraformaldehyde for 2 h at room temperature.

5. Rinse the conduits in PBS thrice for 10 min.

6. Treat the conduits in 1% osmium tetroxide for 1 h.

7. Dehydrate the conduits by ethanol with increased concentrations (25%, 50%, 75%, 95%, and 100%, v/v) subsequently.

8. Coat the dehydrated conduits with platinum.

9. Image the conduits using a SEM (Hitachi S-4800, Japan) (Fig. 2b).

3.4 Dehydration, Swelling, and Degradation Tests

1. Soak the printed alginate conduits without cells in a 4% $CaCl_2$ solution for 30 min.

2. Measure the original conduit weight.

3. Dehydrate the acellular conduits for 4 days at room temperature.

4. Measure the dehydrated conduit weight.

5. Swell and degrade the conduits by soaking the dehydrated conduits for designated period in PBS.

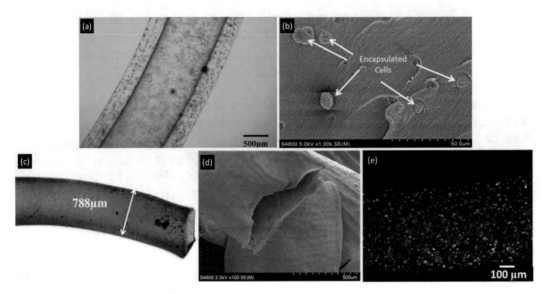

Fig. 2 (**a**) Light microscopic image showing that cells were uniformly encapsulated, and lumen was clearly identified in the center, (**b**) SEM image showing encapsulated cells, (**c**) optical microscope, and (**d**) SEM images of dehydrated 5% alginate conduits, and (**e**) fluorescence image of a conduit cultured for 3 days showing high cell viability (reproduced and adapted from [25])

6. Measure the swollen conduit weight.

7. Calculate the shrinkage rate by weight (SRW) and swelling ratio (SR) using the following equation:

$$SRW = \left(1 - \frac{W_d}{W_O}\right) \times 100\% \qquad (1)$$

$$SR = \frac{W_i - W_d}{W_d} \times 100\% \qquad (2)$$

where W_o is the original conduit weight right after fabrication, W_i is the swollen conduit weight at the predetermined time point, and W_d is the dehydrated conduit weight.

3.5 Dimensional Characterization of Vasculature Conduits During Dehydration, Swelling, and Degradation

1. Observe and measure the conduit dimension throughout the swelling and degradation tests using a light microscope (Motic, BA310, USA) and SEM (Fig. 2c, d, *see* **Note 9**).

2. Calculate the diameter shrinkage rate (DSR) using the following equation:

$$DSR = \left(1 - \frac{D_d}{D_o}\right) \times 100\% \qquad (3)$$

where D_o is the original conduit diameter after fabrication, and D_d is the conduit diameter after dehydration.

3.6 Mechanical Testing

1. Soak the conduits in a 4% $CaCl_2$ solution for 12 h.

2. Cut the conduits to 30 mm long samples.

3. Mount the sample with rectangular mini sandpaper onto a Biotense Perfusion Bioreactor (ADMET, Inc., USA) in order to prevent slippage during testing.

4. Apply the mechanical load onto the conduit until rapture occurs in the middle of the conduit.

5. Record the displacement and load information by a data acquisition system (MTestQuattro System, USA).

6. Calculate the ultimate tensile strength (UTS) from the data.

7. Calculate the estimated burst pressure (BP) using the following equation:

$$BP = 2\frac{UTS \times T}{LD} \qquad (4)$$

where T is the wall thickness (μm) of conduits, and LD is the lumen diameter (i.e., hollow portion, μm).

3.7 Perfusion and Permeability Testing for Acellular Conduit

1. Soak the conduits in a 4% $CaCl_2$ solution for 24 h.

2. Cut the conduits to 8 cm long samples.

3. Customize a media perfusion system using a cell culture media reservoir, a digital pump (Cole-Parmer, USA) and a custom-made perfusion chamber with a clear cover to prevent evaporation.

4. Insert needles into the conduits from two ends, and close the ends using surgery clips to prevent leakage (*see* **Note 10**).

5. Perfuse the cell media through the conduits.

6. Measure the permeability using diffusion rate (μl per hour), which is the volume of media diffused out from conduits to the perfusion chamber in an hour.

3.8 Cell Viability

1. Culture the cellular conduits for 1, 5, and 7 days.

2. Rinse the conduits with PBS twice, 15 min each time.

3. Perfuse the conduits with calcein/ethidium solution to fill it up and immerse the construct in the same solution for 30 min at 37 °C with 5% CO_2.

4. Remove the calcein/ethidium solution.

5. Rinse the conduits with PBS twice, 15 min each time.

6. Observe the conduits using a fluorescence microscope (Leica Microsystems Inc., USA) (Fig. 2e, *see* **Note 11**).

7. Count the red- and green-stained HUVSMCs in each image using ImageJ (National Institutes of Health, USA), and calculate the percentages of viable cells.

3.9 Cell Proliferation Study

1. Culture the cellular conduits for 1, 5, and 7 days.

2. Culture the conduits in 1% (v/v) MTT solution for 4 h at 37 °C and 5% CO_2.

3. Transfer 200 μl (in triplicate) of the supernatant from each well to a 96-well flat-bottom plate.

4. Measure the absorbance of the dye in the conduits at 550 nm using a microplate reader (PowerWave X, Bio-Tek, USA).

3.10 Tissue Histology

1. Culture the cellular conduits for 6 weeks.

2. Fix cells in the strand with 4% paraformaldehyde.

3. Frozen sectioning was used to cut it into 5 μm sections, and hydrate the sample slides following the regular process of histological sample preparation.

4. Stain in Verhoeff's solution for 1 h.

5. Rinse in water twice.

6. Differentiate in 2% ferric chloride for 2 min.

7. Stop differentiation with three changes of DI water.

8. Treat with 5% sodium thiosulfate for 1 min.

9. Wash in running water for 5 min.

10. Counterstain in Van Gieson's solution for 3–5 min.

11. Dehydrate quickly through 95% alcohol, and apply two changes of 100% alcohol.

12. Clear in two changes of xylene for 3 min each.

13. Coverslip with resinous mounting medium.

14. Observe sample slides under a microscope (BX-61, Olympus America, USA) (Fig. 3).

4 Notes

1. Use a low stirring rate due to high concentration of the alginate solution.

2. Coaxial nozzle consists of feed tube, outer tube, and inner tube. Create a hole with the same outer diameter as the feed tube in the barrel of the outer tube to attach the feed tube. Remove the luer lock hub on the barrel of the outer tube using a lathe, and grind the tip to ensure the inner and the outer dispensing tips were even. Align the inner and outer tubes concentrically using a stainless steel fixture manufactured by micromilling, and assemble using laser welding.

3. Dissolve the hematoxylin in alcohol with the aid of gentle heat.

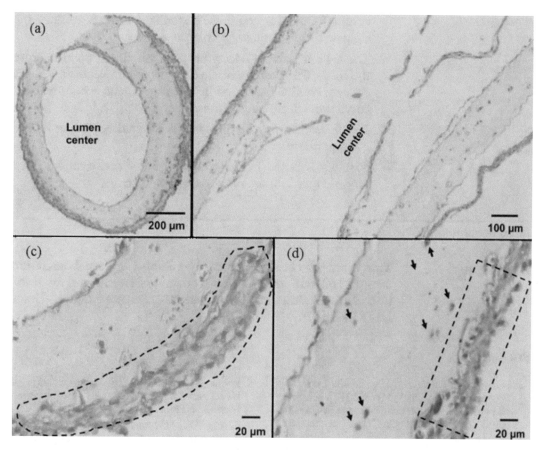

Fig. 3 Histology test for 6-week-cultured conduits: (**a**) reasonable collagen deposition can be observed on long-term-cultured vasculatures, (**b**) delamination of cell sheets observed in some samples during histology sectioning process, and (**c, d**) thick cell sheets were formed on the conduit walls, where arrowheads show intact cells encapsulated in alginate

4. Use 4 ml of DI water to dissolve potassium iodide, and then add iodine. Once iodine is dissolved, dilute this solution by adding 96 ml of DI water. This solution may be prepared fresh as needed or made in larger quantities and stored in brown bottle in the dark at room temperature.

5. Solution should be jet black. Use immediately and discard after use.

6. Use cells within passages 2–5.

7. In order to prevent nozzle clogging caused by alginate, which may crosslink and accumulate at the tip of the nozzle, operator could turn on the syringe pump to keep the CaCl₂ running prior to cell-loaded alginate.

8. Occasionally, the cell-loaded alginate solution may form a big droplet at the nozzle tip without generating continuous

conduit. Operator could use a sterilized tweezer to draw the alginate droplet to initiate conduit extrusion.

9. The 4-hour point is selected as a measurement point because the SR of 5% conduits reached its maximum value at 4 h. 5% alginate conduits are used in this experiment, while their SR curve starts decreasing at the earliest.

10. Needle with a proper diameter is selected according to the conduit diameter.

11. Viability is tested in three conduits. For each sample, three random locations are observed to avoid errors.

Acknowledgments

This work has been supported by National Science Foundation Award #1624515, the National Institutes of Health, and the Institute for Clinical and Translational Science under Grant ULIRR024979.

References

1. Frueh FS, Menger MD, Lindenblatt N, Giovanoli P, Laschke MW (2017) Current and emerging vascularization strategies in skin tissue engineering. Crit Rev Biotechnol 37:613–625

2. Chen M, Przyborowski M, Berthiaume F (2009) Stem cells for skin tissue engineering and wound healing. Crit Rev Biomed Eng 37:399–421

3. Wu Y, Wong YS, Fuh JYH (2017) Degradation behaviors of geometric cues and mechanical properties in a 3D scaffold for tendon repair. J Biomed Mater Res A 105:1138–1149

4. Wu Y, Fuh JYH, Wong YS, Sun J (2015) Fabrication of 3D scaffolds via E-jet printing for tendon tissue repair. In: Proc. ASME 10th int manuf sci eng conf 2, pp 3–8. https://doi.org/10.1115/MSEC20159367

5. Palakkan AA, Hay DC, Pr AK, Tv K, Ross JA (2013) Liver tissue engineering and cell sources: issues and challenges. Liver Int 33:666–676

6. Batchelder CA, Martinez ML, Tarantal AF (2015) Natural scaffolds for renal differentiation of human embryonic stem cells for kidney tissue engineering. PLoS One 10:1–18

7. Sheridan WS, Duffy GP, Murphy BP (2012) Mechanical characterization of a customized decellularized scaffold for vascular tissue engineering. J Mech Behav Biomed Mater 8:58–70

8. Pober JS, Tellides G (2012) Participation of blood vessel cells in human adaptive immune responses. Trends Immunol 33:49–57

9. Chang WG, Niklason LE (2017) A short discourse on vascular tissue engineering. Regen Med 2:7

10. Ravi S, Chaikof E (2010) Biomaterials for vascular tissue engineering. Regen Med 5:1–21

11. McKee JA, Banik SSR, Boyer MJ, Hamad NM, Lawson JH, Niklason LE et al (2013) Human arteries engineered in vitro. EMBO Rep 4:633–638

12. Watson T, Pope A, van Pelt N, Ruygrok PN (2015) Evaluation of previously cannulated radial arteries as patent coronary artery bypass conduits. Tex Heart Inst J 42:448–449

13. Kannan RY, Salacinski HJ, Butler PE, Hamilton G, Seifalian AM (2005) Current status of prosthetic bypass grafts: a review. J Biomed Mater Res B 74:570–581

14. Greisler HP (1990) Interactions at the blood/material Interface. Ann Vasc Surg 4:98–103

15. Dahl SLM, Koh J, Prabhakar V, Niklason LE (2003) Decellularized native and engineered arterial scaffolds for transplantation. Cell Transplant 12:659–666

16. Canham PB, Talman EA, Finlay HM, Dixon JG (1991) Medial collagen organization in human arteries of the heart and brain by

polarized light microscopy. Connect Tissue Res 26:121–134

17. L'Heureux N, Pâquet S, Labbé R, Germain L, Auger FAA (1998) Completely biological tissue-engineered human blood vessel. FASEB J 12:47–56

18. Caves JM, Kumar VA, Martinez AW, Kim J, Ripberger CM, Haller CA et al (2010) The use of microfiber composites of elastin-like protein matrix reinforced with synthetic collagen in the design of vascular grafts. Biomaterials 31:7175–7182

19. Salerno A, Zeppetelli S, Di Maio E, Iannace S, Netti PA (2011) Processing/structure/property relationship of multi-scaled PCL and PCL-HA composite scaffolds prepared via gas foaming and NaCl reverse templating. Biotechnol Bioeng 108:963–976

20. Norotte C, Marga F, Niklason L, Forgacs G (2010) Scaffold-free vascular tissue engineering using bioprinting. Biomaterials 30:5910–5917

21. Xu C, Chai W, Huang Y, Markwald RR (2012) Scaffold-free inkjet printing of three-dimensional zigzag cellular tubes. Biotechnol Bioeng 109:3152–3160

22. Seliktar D, Dikovsky D, Napadensky E (2013) Bioprinting and tissue engineering: recent advances and future perspectives. Isr J Chem 53:795–804

23. Ozbolat IT, Yu Y (2013) Bioprinting toward organ fabrication: challenges and future trends. IEEE Trans Biomed Eng 60:691–699

24. Datta P, Ayan B, Ozbolat IT (2017) Bioprinting for vascular and vascularized tissue biofabrication. Acta Biomater 51:1–20. https://doi.org/10.1016/j.actbio.2017.01.035

25. Zhang Y, Yu Y, Akkouch A, Dababneh A, Dolati F, Ozbolat I (2015) In vitro study of directly bioprinted perfusable vasculature conduits. Biomater Sci 3:134–143

26. Yu Y, Moncai K, Li J, Peng W, Rivero I, Martin JA, Ozbolat I (2016) Three-dimensional bioprinting using self-assembling scalable scaffold-free 'tissue strands' as a new bioink. Sci Rep 6:28714

27. Akkouch A, Yu Y, Ozbolat IT (2015) Microfabrication of scaffold-free tissue strands for three-dimensional tissue engineering. Biofabrication 7(3):031002

Chapter 12

Principles of Spheroid Preparation for Creation of 3D Cardiac Tissue Using Biomaterial-Free Bioprinting

Chin Siang Ong, Isaree Pitaktong, and Narutoshi Hibino

Abstract

Biomaterial-free three-dimensional (3D) bioprinting is a relatively new field within 3D bioprinting, where 3D tissues are created from the fusion of 3D multicellular spheroids, without requiring biomaterial. This is in contrast to traditional 3D bioprinting, which requires biomaterials to carry the cells to be bioprinted, such as a hydrogel or decellularized extracellular matrix. Here, we discuss principles of spheroid preparation for biomaterial-free 3D bioprinting of cardiac tissue. In addition, we discuss principles of using spheroids as building blocks in biomaterial-free 3D bioprinting, including spheroid dislodgement, spheroid transfer, and spheroid fusion. These principles are important considerations, to create the next generation of biomaterial-free spheroid-based 3D bioprinters.

Key words Tissue engineering, Biofabrication, 3D bioprinting, Biomaterial-free, Spheroids

1 Introduction

Three-dimensional (3D) cell culture and the creation of 3D multicellular spheroids have been described for almost five decades, but it is only recently that spheroids have been employed more routinely for the purpose of tissue engineering [1] and tissue regeneration [2]. 3D multicellular spheroids offer the opportunity to create 3D tissue without the use of biomaterials. Advances in 3D bioprinting technology have led to the new field of biomaterial-free 3D bioprinting [3–8].

Traditional biomaterial-dependent 3D bioprinting [9] has limitations, such as the intrinsic stress to live cells by heat exposure or shear stress during 3D bioprinting, isolation of 3D bioprinted cells from their neighbors, and the risk of foreign material in the biomaterials, which may be relatively contraindicated for eventual intended clinical use. In contrast, using spheroids as building blocks for 3D bioprinting is gentler and may recapitulate the developmental biological environment encountered early on, in the developing embryo.

Jeremy M. Crook (ed.), *3D Bioprinting: Principles and Protocols*, Methods in Molecular Biology, vol. 2140,
https://doi.org/10.1007/978-1-0716-0520-2_12, © Springer Science+Business Media, LLC, part of Springer Nature 2020

In this chapter, we discuss the principles and important considerations for creation of 3D cardiac tissue from multicellular spheroids using biomaterial-free 3D bioprinting based on our experience with the Regenova 3D bioprinter (Cyfuse Inc., Tokyo, Japan). We divide the protocol and principles into four distinct steps: spheroid optimization (**step 1**), spheroid dislodgement (**step 2**), spheroid transfer (**step 3**), and spheroid fusion (**step 4**).

2   Materials

2.1   Cell Source and Media

2.1.1   Induced Pluripotent Stem Cells (iPSCs) and Differentiation to Cardiac Lineage

1. iPSCs.
   - In-house-derived iPSCs reprogrammed from the peripheral blood mononuclear cells of a healthy human donor.
   - Sourced from another institution.
   - Commercially purchased.

2. Cytokines or chemicals for iPSC differentiation to cardiomyocytes.
   - Dependent on intended cell type post differentiation.
   - For cardiac differentiation, we use
     - Y27632 ROCK inhibitor (Tocris, cat. no. 1254).
     - CHIR99021 (Tocris, cat. no. 4423).
     - IWR-1 (Sigma, cat. no. I0161).

3. Medium for iPSC/iPSC-derived cardiomyocytes.
   - For iPSCs, we use E8 media (Invitrogen cat. no. A1517001).
   - For human iPSC-derived cardiomyocytes (hiPSC-CMs), we use RPMI media with B27 without insulin supplement, until the hiPSC-CMs start to beat. Then the media is switched to RPMI media with B27 with insulin supplement:
     - RPMI 1640 with 2.5 mM Glutamine (Invitrogen, cat. no. 11875093).
     - B27 supplement without insulin (Invitrogen, cat. no. A1895601).
     - B27 supplement with insulin (Invitrogen, cat. no. 17504044).

4. Culture plate and coating.
   - Geltrex LDEV-free hESC-qualified (Invitrogen, cat. no. A1413202).
   - 6-Well cell culture plates (Corning, cat. no. 3516).

2.1.2 *Commercial Cell Lines and Culture Media*

1. Human umbilical vein endothelial cells (HUVECs) (Lonza, cat. no. CC-2935).

2. HUVEC media EGM-Plus BulletKit (Lonza, cat. no. CC-5035).

3. Human adult cardiac fibroblasts (FBs) (ScienCell, cat. no. 6330).

4. Fibroblast medium FM-2 (ScienCell, cat. no. 2331).

2.2 Cell Dissociation

1. Cells.

2. Cell media.

3. T-175 cell culture flasks.

4. 6-Well cell culture plates.

5. Pipette set.

6. Pipette tips.

7. Motorized pipette controller.

8. Serological pipettes.

9. Conical tubes (50 ml).

10. Microcentrifuge tubes.

11. Optical microscope.

12. Penicillin/streptomycin (P/S) solution or Normocin (Invivogen, cat. no. ant-nr-2).

13. TrypLE Express enzyme (1×), no phenol red (Gibco, cat. no. 12604013).

14. Cell scrapers.

15. Trypan blue solution, 0.4%.

16. Manual hemocytometer.

17. Automated cell counter (optional) (Countess II, Thermo Fisher, cat. no. AMQAX1000).

2.3 Spheroid Creation

1. Ultralow attachment 96-well U-bottom plates (PrimeSurface, S-Bio, cat. no. MS-9096UZ).

2. Multichannel pipette.

3. Pipetting reservoir/tray.

4. Pipette tips—normal and multichannel.

5. Conical tubes (50 ml).

6. Sterile cell culture container that can hold more than 200 ml, e.g., T-175 cell culture flask.

3   Methods

3.1   Spheroid Optimization/Creation

1. Principles (*see* **Note 1**).

2. Before isolating cells, determine the cell types and cell type ratio/composition of the spheroids, then determine the size, thickness, shape, and quantity of 3D cell constructs desired. Next, determine the number of experimental groups and implement modifications for the experimental group(s) as required, e.g., addition of growth factors to the cell media (*see* **Note 2**).

3. Calculate the approximate number of spheroids and cells needed, taking into account a 10% loss during fluid transfer and a 20% loss during 3D bioprinting (*see* **Note 3**, Table 1).

4. Using the number of required hiPSC-CMs, determine the number of wells of hiPSC-CMs that are needed to make the 3D construct(s). To ensure consistent spheroid quality and the 3D constructs, visually grade hiPSC-CMs using a light microscope (*see* **Note 4**).

5. Isolate hiPSC-CMs using TrypLE (*see* **Note 5**).

6. Remove the cells from the incubator and neutralize TrypLE using 2 ml of RPMI/B27 supplement media in each well at room temperature (*see* **Note 6**). Observe cardiomyocytes under an optical microscope to check for cell dissociation.

7. Pool the isolated cardiomyocytes from the respective wells and transfer them into one 50 ml conical tube using a motorized pipette controller (*see* **Note 7**).

8. Centrifuge the cell suspension for 5 min at $250 \times g$ at room temperature to obtain a pellet (*see* **Note 8**).

9. Resuspend the pellet in 10 ml of RPMI cell media supplemented with B-27 with insulin (RPMI/B-27 cell media) (*see* **Note 9**).

10. To count the cells, pipette 10 μl of cells in cell media and stain with an equal amount of 0.4% Trypan Blue solution (*see* **Note 10**).

11. Use a manual hemocytometer to count and obtain the concentration and cell viability of the cell suspension (*see* **Note 11**).

12. Note the number and percentage of live cells and the concentration of the cell suspension, and verify that there is a sufficient number of cardiomyocytes to make the desired number of spheroids (*see* **Note 12**).

13. Repeat **steps 5–12** for the isolation of FBs and ECs (*see* **Note 13**).

Table 1
Example spheroid optimization with three experimental groups to make 100 spheroids per group

Group	Number of CMs required	Number of FBs required	Number of HUVECs required	Number of spheroids/cells required
1: CM:FB:EC 70:15:15	0.70 * 100 * 33,000 = 2,310,000 cells	0.15 * 100 * 33,000 = 990,000	0.15 * 100 * 33,000 = 990,000	81 * 120% ≈ 100 spheroids
2: CM:FB:EC 70:0:30	0.70 * 100 * 33,000 = 2,310,000 cells	0	0.30 * 100 * 33,000 = 990,000	81 * 120% ≈ 100 spheroids
3: CM:FB:EC 70:30:0	0.70 * 100 * 33,000 = 2,310,000 cells	0.30 * 100 * 33,000 = 990,000	0	81 * 120% ≈ 100 spheroids
Total cells required (considering 10% fluid transfer loss)	2,310,000 * 3 * 1.1 = 7,623,000 cells	990,000 * 2 * 1.1 = 1,633,500 cells	990,000 * 2 * 1.1 = 1,633,500 cells	7,623,000 + 1,633,500 + 163,500 = 10,890,000 cells

Table 2
Example calculation of mixed cell suspension, or cell "cocktail" to make 100 spheroids (CM:FB:EC 70:15:15—Experimental Group 1)

Number of spheroids	Total volume	CM volume	FB volume	EC volume	Media volume
100 (no fluid transfer losses)	200 µl $*$ 100 = 20 ml	$2{,}310{,}000/10^6$ = 2.31 ml	$990{,}000/10^6$ = 0.99 ml	$990{,}000/10^6$ = 0.99 ml	$20 - 2.31 - 0.99 - 0.99$ = 15.71 ml
100 (10% fluid transfer losses)	$20 * 1.1$ = 22 ml	$2.31 * 1.1$ = 2.541 ml	$0.99 * 1.1$ = 1.089 ml	$0.99 * 1.1$ = 1.089 ml	$15.71 * 1.1$ = 17.281 ml

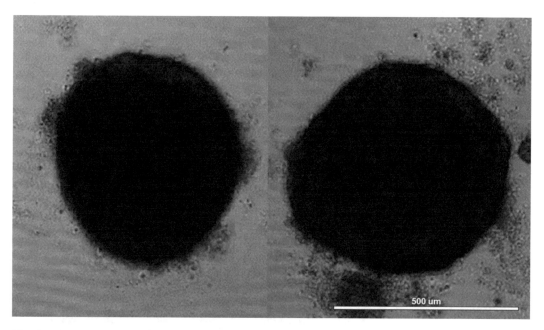

Fig. 1 The cells in each well should aggregate at the bottom of the ultralow attachment 96-well U-bottom plates to form spheroids by gravity. These spheroids can be visually inspected under light microscopy, where they should appear circular by two-dimensional projection. (Reproduced from [3] with permission from JoVE)

14. Verify that there is a large enough volume of FBs, HUVECs, and CMs to make the desired number of spheroids. Note down the volume of CMs, FBs, ECs, and additional cell media (RPMI/B27) needed (*see* **Note 14**, Table 2).

15. Mix the calculated volumes of each cell type (CMs, FBs, and ECs) in RPMI/B27 cell media to generate the mixed cell suspension in a 50-ml conical tube (*see* **Note 15**).

16. Using a multichannel pipette filler, set at 200 µl, distribute the mixed cell suspension into each column of ultralow attachment 96-well U-bottom plates (*see* **Note 16**).

17. Incubate the 96-well plates for 3 days (37 °C, 5% carbon dioxide, 95% humidity) (*see* **Note 17**).

18. Inspect for the presence of mixed cell multicellular spheroids at the center and bottom of individual wells, under light microscopy (*see* **Note 18**, Fig. 1).

3.2 Spheroid Dislodgement

The principle behind spheroid dislodgement is that the spheroid created should be as easy to pick up as possible, minimizing damage to the shape of spheroids during the process and troubleshooting. The following list includes what we routinely consider when seeking to establish the optimal conditions for the dislodgement of spheroids.

- Spheroid culture conditions: We seek to optimize the spheroid culture conditions by modifying parameters as documented in **Note 1**. In addition, the type of cells (adult or neonatal), passage number, timing of harvesting cells, culture duration, and culture medium (e.g., addition of growth factor) also affect the adhesion of spheroids to the bottom of the well.

- Cell types within the spheroids: Cell types are inherently different, and some cell types may secrete more extracellular matrix or be more "sticky" or attach to the bottom of the well more than others. Reduction in the proportion of fibroblasts in the spheroid, in particular, can decrease the adhesion of spheroids to the bottom of the well. If a specific type of tissue with a specific cell type ratio is desired to answer a scientific question, the component spheroid is inherently more adherent to the bottom of the well.

- Mechanical force (such as tapping, hitting) may be required to dislodge the spheroid from the bottom of the well.

- ECM-secreting cells: For cells that do not adhere well to each other, supporting cells that secrete ECM, such as fibroblasts, or even biomaterial, may be needed for spheroid creation.

3.3 Spheroid Transfer

There are several methods for transferring spheroids, such as by machine-generated mechanical suction or manual pipetting by hand. The principle behind spheroid transfer is that the spheroids need to be sufficiently compact in order to be picked up and moved during bioprinting. The way to increase spheroid compaction is generally the opposite to that discussed in the previous section. For example, increasing culture duration and fibroblast cell ratio will lead to more compact spheroids at the cost of the spheroids being more attached to the bottom of the wells. This, however, is not always the case as the level of compaction depends largely on the shape in which the spheroids form. While preferable that spheroids form as spheres and gradually compact, they may be more elliptical or "pancake"-like, especially if they attach to the bottom of the well in the early stages of spheroid formation (first 24 h). Spheroid reoptimization may be needed if this step fails on multiple occasions, and there is an unacceptable proportion of spheroid losses.

3.4 Spheroid Fusion

The spheroids need to fuse together in order to form an organized 3D tissue. Spheroid size is one of the most important factors for the fusion of spheroids. The principles behind spheroid fusion are as follows:

- There must be sufficient overlap of spheroids of at least 50 μm (Fig. 2) to begin the fusion process.

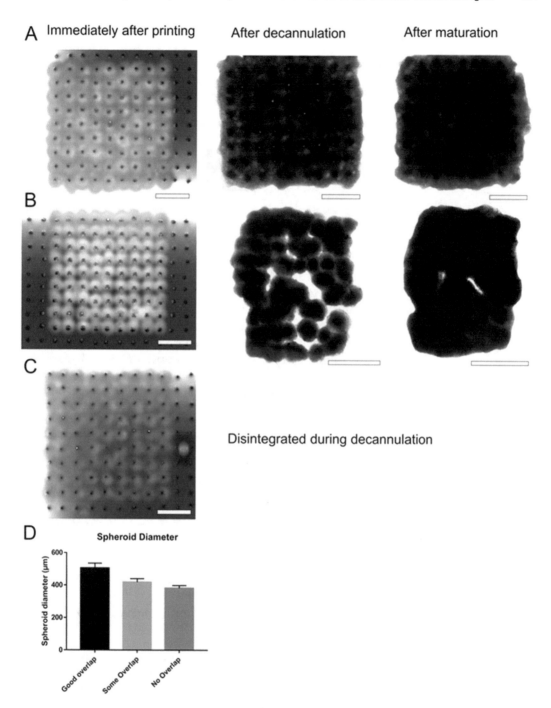

Fig. 2 Fusion of spheroids: Degree of spheroid overlap determines the structural properties of 3D bioprinted tissue construct. (**a–c**) Representative images of 3D bioprinted patches (CM:FB:EC 70:15:15) with (**a**) spheroid overlap of at least 50 μm, (**b**) some overlap, 1–50 μm, and (**c**) no overlap; each immediately after bioprinting, after decannulation, and after maturation. Scale bar: 1000 μm. (Reproduced from [4] with permission from Scientific Reports)

- If the spheroids are too large and there is too much overlap, adjacent spheroids may push each other away, resulting in the creation of disorganized and fragile tissue.

- There must be sufficient time for spheroids to fuse (we culture for 3 days).

- The fusion process can be impaired if there are misplaced or lost spheroids during 3D bioprinting.

- As the 3D bioprinting process is not completely sterile, we recommend the addition of antibiotics immediately after 3D bioprinting, such as penicillin/streptomycin or Normocin to reduce contamination during the fusion process. Antibiotics may be added during spheroid creation, if there are significant issues with contamination despite maximal precautions and the addition of antibiotics immediately after 3D bioprinting. We add Normocin at 1:500 to our cell media. However, the addition of antibiotics to cell culture media before 3D bioprinting and for long durations after the tissue has been created and is stable (not disintegrating) is not recommended as this may lead to antibiotic resistant bacteria, and long-term Normocin exposure may harm the cells. The required accessories of the 3D bioprinter should be assembled before autoclaving to minimize handling after autoclaving for improved sterility.

4   Notes

1. The principle behind spheroid optimization is to create a 3D multicellular spheroid that is

   - Not too adherent to the bottom of the ultralow attachment U-well, that will result in significant effort during spheroid dislodgement (**step 2**).

   - Compact and strong enough to prevent disintegration during spheroid transfer (**step 3**).

   - Appropriately sized to ensure adequate fusion between spheroids (**step 4**, Fig. 2).

   The following parameters (nonexhaustive list) may be adjusted to obtain the optimized spheroid. Our parameters [4] are stated in parentheses.

   - Type of cells (hiPSC-CMs, FBs, HUVECs).

   - Cell ratio (CM:FB:EC 70:15:15).

   - Number of cells seeded into 96-well plate for one spheroid (33,000 cells).

   - Duration of 3D culture (72 h).

The characteristics of an optimized spheroid vary widely. For us, these are the characteristics of the spheroid that we consider acceptable for bioprinting:

- Resultant spheroid diameter (500–600 μm).
- Resultant cell viability within spheroid (93.3%).
- Spheroid circularity/roundness (70%).
- Spheroid smoothness (70%).

2. Examples of growth factors include fibroblast growth factor-2 (FGF-2), vascular endothelial growth factor (VEGF), and thyroxine.

3. Determine the desired spheroid composition/cell type ratio, e.g., CM:FB:HUVEC 70:15:15. Also, determine the total number of spheroids desired for each ratio (*see* Table 1). It is important to consider the loss of spheroids between steps. Excessive loss will result in decreased number of spheroids that can be used to eventually print the 3D tissue constructs. The percentage of spheroid loss varies widely, depending on spheroid optimization as discussed in **Note 1** and can range from 10% to as high as 90%, depending on the acceptable threshold for spheroid loss, the desire to create the tissue, and the perseverance of the tissue engineer.

To calculate the total cells needed per group, multiply the number of spheroids by 33,000 (the number of cells per well, or spheroid) and factor in an extra 10% for fluid transfer losses to obtain the total number of cells needed. For example, total cells per group = 100 spheroids × 33,000 cells/well × 1.1 transfer loss adjustment.

Calculate the total number of cells, and total number of cells of each type, considering the number of experimental groups.

The following table shows the calculations of cells required to create 100 spheroids per group, considering 20% losses during transfer with CM:FB:EC cell type ratios of 70:15:15, 70:0:30, and 70:30:0, respectively. Note that each well in the 96-well plates used for bioprinting holds 200 μl, and we create spheroids using 33,000 cells per spheroid, resulting in a concentration of 165,000 cells per ml.

4. Electrophysiological tests and immunostaining can also be performed on the hiPSC-CMs as 2D monolayers for quality control and to ensure that hiPSC-CMs are differentiated, functional, and able to form electrical connections in vitro. Grading scale [4]: connected beating sheet (Grade A), connected beating areas (Grade B), isolated beating areas (Grade C), and not beating (Grade D). We culture our hiPSC-CMs in 6-well plates. Depending on the culture

conditions, seeding density of the IPSCs, the confluency when differentiation factors are added, and the overall care and attention during cell feeding (e.g., care taken not to wash the cells off as a result of using high speed of cell media ejection by the pipette filler), the resultant hiPSC-CMs obtained per well and thus the number of wells needed may vary.

5. The hiPSC-CMs should ideally receive a change of media 1 h before isolation and be washed once with DPBS before the addition of TrypLE (1.5 ml each well for 4 min 30 s at 37 °C). An alternative to TrypLE is trypsin/EDTA 0.05% (2 ml in each well for 5 min at room temperature). Use hiPSC-CMs from regions of 2D cultures that were beating to create spheroids. The nonbeating areas of the 2D monolayer in the well can be marked under the culture plates and manually removed by scaping, prior to cell dissociation.

6. If using trypsin, neutralize trypsin using 2 ml of trypsin inhibitor.

7. Using a 1 ml pipette or a motorized pipette controller, spray each well until the cells come off (this will be visibly apparent). Alternatively, a cell scraper can be used to scrape cells from bottom of well gently and with minimal motion, although this may decrease cell viability. This scraping method is suggested only for situations where the cells are very adherent. The duration of exposure to the cell dissociation agent may be prolonged, again at the cost of cell viability.

8. If floating "strings" of cells are present after spinning (likely dead cells and cell debris), remove them and spin down again.

9. If clumps are visible, resuspend the cells in a small amount of media first, using a 1-ml hand pipette to pipette up and down, up to 20 times.

10. Make sure the contents of the conical tube are well mixed, by shaking, rotating, or pipetting several times using a motorized pipette controller, before drawing any amount to ensure an even distribution of cells in the conical tube. This is because cells have a propensity to settle at the bottom due to gravity. However, avoid using a vortex machine, as this will decrease cell viability. It is important to use the 20 μl pipette when transferring cells because the tip for the 10 μl pipette is very small and can damage the cells due to shearing, artificially increasing the percentage of dead cells in the sample. When using a pipette to draw a sample of the cells, draw from the middle of the conical tube containing the cell suspension (avoid pressing the tip against the side of the conical tube as this may prevent cells from entering and being drawn). If unsure, this cell counting process may be repeated till a consistent result is obtained. It is especially important as an error in

this step will affect spheroid characteristics, in particular diameter, and all subsequent steps, as well as the characteristics of the 3D bioprinted constructs.

11. We use a manual cell counter to count CMs, rather than an automatic cell counter, which may potentially count Geltrex fragments as cells, and clumped CMs as a single cell. If clumps of hiPSC-CMs are seen in this step, we suggest returning to **step 9** and repeating this process to get an accurate cell count. Estimate the number of cells from the total number of wells of CMs used, based on the previous experience. If the counted number is drastically different than the estimated amount, perform a recount (restarting at **step 10**). Dilute with more cell media if the cell density is too high, which may affect the cell count.

12. At this step, CMs may be fixed for flow cytometry for cardiac specific antibodies, if additional cell quality checks are desired.

13. Instead of manual cell counting, an automated cell counter may be used for FBs and ECs as we grow these cells in culture flasks without any coating or Geltrex. The cell dissociation steps (duration of incubation in cell dissociation agent, temperature, type of cell dissociation agent) should be optimized for each cell line and is dependent on culture duration, confluency, and cell passage. This process may be done in parallel with an assistant simultaneously as CMs are being isolated to ensure the highest cell viability of all cell types, just prior to mixing.

14. Once the necessary number of cells of each type are isolated and the concentration (cells/ml) of each cell suspension is determined, calculate the "cocktail" of cells for each group (volume of individual cell suspensions and media to obtain the desired ratio, for a total of 200 µl per well). We demonstrate an example calculation of the volume of cell suspensions required, assuming all the CM, FB, and EC cell suspensions have 10^6 live cells/ml (Table 2). Media is then added to a total volume of 200 µl per well.

15. Other cell ratios may also be used, e.g., experimental groups 2 and 3 in Table 1.

16. Transfer the cell suspension into a reservoir for easy access, mixing the cell suspension in the reservoir each time before using a multichannel pipette filler to transfer the contents to the 96-well plates, to ensure that the mixed cell suspension is homogeneous. If bubbles appear in the pipette tips, change the tip. This is likely due to cell debris, and the presence of bubbles will alter the volume and number of cells delivered into each well. Additionally, bubbles in the wells may result in cell loss. The benefit of taking 10% fluid transfer losses into account

earlier is apparent in this step, because if the cell suspension volume is calculated exactly, during the pipetting of cells into the final column of the plate, bubbles are very likely to form, resulting in small spheroids. To prevent cell wastage, this 10% may be reduced to 5%, depending on experience.

17. During this time, the sequence of spheroid formation may be observed under an optical microscope [4].

18. In a 2D microscopic projection, these spheroids will appear circular in shape (Fig. 1) [3]. However, the spheroids may actually be elliptical in shape. After cell seeding into ultralow attachment 96-well plates, spheroids should form within 24 h and should start beating within 48 h. Spheroids that are not beating should not be used for 3D bioprinting to ensure that the final 3D bioprinted cardiac patch is functional. In addition, the formation of irregularly shaped spheroids, from our experience, may herald infection and contamination, or the overgrowth of fibroblasts or an undifferentiated iPSC-derived cell type. If contamination is suspected, the entire plate should not be used and the spheroid creation process should be restarted from the beginning. Immunofluorescence staining of spheroids for connexin 43 and troponin may be used to assess spheroid cellular composition and quality [3].

5   Conclusion

This chapter is an updated laboratory protocol, modified from our previous publication [4] and following 2 years' experience of biomaterial-free spheroid-based 3D bioprinting. We have sought to document tips and advice, discovered from repeated experimental trial and error. The chapter seeks to highlight the importance of the 3D cell spheroid as a transient building block, as source cells for 3D bioprinting into a final 3D bioprinted tissue construct.

We suggest that the best process of tissue creation using biomaterial-free spheroid-based 3D bioprinting is based on the dynamic balance and flexible interactions of the four steps we have outlined. We use the Goldilocks principle in pursuit of the ideal spheroid for 3D bioprinting (not too small, not too large, not too compact, not too loose, etc.), while recognizing the inherent limitations of 3D culture and the 3D bioprinter and the cell types to be bioprinted. This conceptual breakdown serves as a guide to understand the principles behind biomaterial-free spheroid-based 3D bioprinting and to develop the next generation of 3D bioprinters.

Acknowledgments

The protocol and content of this chapter is based mostly on the published work by the authors [3, 4]. Written reprint permission has been obtained from the Journal of Visualized Experiments [3] and Nature Publishing Group [4], respectively, and is available from the editors upon request. In addition, we would like to acknowledge the other members in the laboratory who have used the 3D bioprinter, namely, Tom Zhang, Takuma Fukunishi, Cecilia Lui, Enoch Yeung, Yang Bai, and Joseph Boktor, and may have shared helpful experience.

References

1. Laschke MW, Menger MD (2016) Life is 3D: boosting spheroid function for tissue engineering. Trends Biotechnol 35(2):133–144. https://doi.org/10.1016/j.tibtech.2016.08.004

2. Ong CS, Zhou X, Han J, Huang CY, Nashed A, Khatri S et al (2018) In vivo therapeutic applications of cell spheroids. Biotechnol Adv 36 (2):494–505. https://doi.org/10.1016/j.biotechadv.2018.02.003

3. Ong CS, Fukunishi T, Nashed A, Blazeski A, Zhang H, Hardy S et al (2017) Creation of cardiac tissue exhibiting mechanical integration of spheroids using 3D bioprinting. J Vis Exp (125):e55438. https://doi.org/10.3791/55438

4. Ong CS, Fukunishi T, Zhang H, Huang CY, Nashed A, Blazeski A et al (2017) Biomaterial-free three-dimensional bioprinting of cardiac tissue using human induced pluripotent stem cell derived cardiomyocytes. Sci Rep 7:4566

5. Itoh M, Nakayama K, Noguchi R, Kamohara K, Furukawa K, Uchihashi K (2015) Scaffold-free tubular tissues created by a bio-3D printer undergo remodeling and endothelialization when implanted in rat aortae. PLoS One 10(9): e0136681. https://doi.org/10.1371/journal.pone.0136681

6. Yurie H, Ikeguchi R, Aoyama T, Kaizawa Y, Tajino J, Ito A et al (2017) The efficacy of a scaffold-free Bio 3D conduit developed from human fibroblasts on peripheral nerve regeneration in a rat sciatic nerve model. PLoS One 12 (2):e0171448. https://doi.org/10.1371/journal.pone.0171448

7. Kizawa H, Nagao E, Shimamura M, Zhang G, Torii H (2017) Scaffold-free 3D bio-printed human liver tissue stably maintains metabolic functions useful for drug discovery. Biochem Biophys Rep 10:186–191. https://doi.org/10.1016/j.bbrep.2017.04.004

8. Moldovan L, Barnard A, Gil CH, Lin Y, Grant MB, Yoder MC et al (2017) iPSC-derived vascular cell spheroids as building blocks for scaffold-free biofabrication. Biotechnol J 12:17004444. https://doi.org/10.1002/biot.201700444

9. Moldovan NI, Hibino N, Nakayama K (2017) Principles of the Kenzan method for robotic cell spheroid-based three-dimensional bioprinting. Tissue Eng Part B Rev 23(3):237–244. https://doi.org/10.1089/ten.TEB.2016.0322

Chapter 13

Bioprinting for Human Respiratory and Gastrointestinal In Vitro Models

Manuela Estermann, Christoph Bisig, Dedy Septiadi, Alke Petri-Fink, and Barbara Rothen-Rutishauser

Abstract

Increasing ethical and biological concerns require a paradigm shift toward animal-free testing strategies for drug testing and hazard assessments. To this end, the application of bioprinting technology in the field of biomedicine is driving a rapid progress in tissue engineering. In particular, standardized and reproducible in vitro models produced by three-dimensional (3D) bioprinting technique represent a possible alternative to animal models, enabling in vitro studies relevant to in vivo conditions. The innovative approach of 3D bioprinting allows a spatially controlled deposition of cells and biomaterial in a layer-by-layer fashion providing a platform for engineering reproducible models. However, despite the promising and revolutionizing character of 3D bioprinting technology, standardized protocols providing detailed instructions are lacking. Here, we provide a protocol for the automatized printing of simple alveolar, bronchial, and intestine epithelial cell layers as the basis for more complex respiratory and gastrointestinal tissue models. Such systems will be useful for high-throughput toxicity screening and drug efficacy evaluation.

Key words In vitro cultures, Alveolar epithelial cells, Bronchial epithelial cells, Intestine epithelial cells, Bioprinting technique

1 Introduction

Epithelial tissues like the skin and respiratory and gastrointestinal tract act as primary structural barriers and the first line of interaction with drugs or chemicals [1]. Conventional approaches to assessing potential adverse effects are mainly based on animal studies [2], which have largely contributed to understanding pathological processes and the safety of drugs and chemicals [3, 4]. However, humans and animals differ in physiology, metabolic processes, and disease progression, leading to increased interest in animal-free in vitro alternatives [4, 5]. Therefore, the development of relevant, reliable, and predictive in vitro models that mimic the complex structure and function of human tissues are needed. To date, only validated 3D dermal models exist for toxicity

Jeremy M. Crook (ed.), *3D Bioprinting: Principles and Protocols*, Methods in Molecular Biology, vol. 2140,
https://doi.org/10.1007/978-1-0716-0520-2_13, © Springer Science+Business Media, LLC, part of Springer Nature 2020

testing of irritant chemicals (e.g., OECD Test Guideline 439) [6], while other barrier systems such as lung or intestine are lacking.

Recent tissue engineering strategies have promoted a combination of 3D printing technology and biological inks (i.e., hydrogels/biomaterials) with cells to produce standardized and reproducible cell models for high-throughput screening [7]. This platform enables spatial control of the deposition of biomaterial and cells for the generation of complex 3D structures in a layer-by-layer fashion [8]. Bioprinting is being applied in many different fields, including regenerative medicine through personalized tissue and organ engineering [7], as well as 3D cancer modeling for investigation of disease progression and as a platform for drug screening [9, 10]. As an example, a 3D lung model composed of Matrigel™ as extracellular matrix, human epithelial cell line (i.e., A549) and endothelial cell line (i.e., EA.hy926) was constructed using 3D bioprinting [10].

Despite the potential for bioprinting, standardized protocols that facilitate the development of more complex models are lacking. Herein, we provide a detailed protocol using a widely available bioprinting device to produce simple epithelial cell layers as the basis for more complex respiratory and gastrointestinal in vitro models. The protocol has been applied to create different human epithelial cell layers using the alveolar epithelial cell line A549, the bronchial epithelial cell line 16HBE14o-, and the intestine epithelial cell line Caco-2 and is compatible with any other epithelial cells from internal barriers such as liver or kidney, as well as meso- and endothelial cells. The bioprinting platform represents an advance toward manually seeded cell models since it allows automatized and spatially controlled deposition for the reproducible and high-throughput creation of complex models. Additionally, the protocol addresses several technical issues that may be encountered during the printing process.

2   Materials

2.1   General Materials

1. Sterile biosafety cabinet equipped with laminar flow.
2. Incubator (37 °C, 5% CO_2).
3. Water bath.
4. 15 and 50 mL tubes.
5. Glass pipettes for aspiration.
6. Serological pipettes.
7. Centrifuge (e.g., Centrifuge 5702 from Eppendorf).
8. Cell counter.
9. Protective gloves, lab coat, and safety goggles.

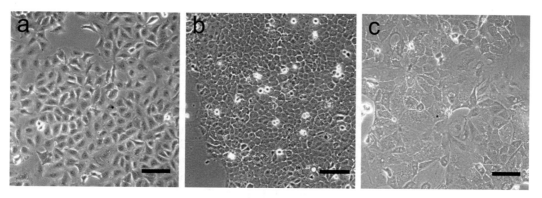

Fig. 1 Light microscope images of cell lines (**a**) A549, (**b**) 16HBE14o-, and (**c**) Caco-2 grown in culture flasks. Scale bar = 100 μm

2.2 Cell Lines

1. A549: lung epithelial cell line (Fig. 1a).

   A549 is a type II alveolar epithelial cell line originating from a human lung adenocarcinoma [11]. Confluent cells show characteristic features of type II cells including lamellar bodies [12, 13]. A549 cells are extensively referenced in the literature related to hazard assessment of aerosol and nanoparticle exposure [10, 14]. The cell line is available from the American Type Culture Collection (ATCC® CCL-185™, LGC Standards, Germany).

2. 16HBE14o-: bronchial epithelial cell line (Fig. 1b).

   16HBE14o- is an immortalized SV40 large T-antigen transformed bronchial epithelial cell line, originating from a human heart-lung transplant patient [15, 16]. 16HBE14o- cells form confluent monolayers and express extensive tight junctions and are therefore suitable for modeling the barrier properties of the human bronchial epithelium in vitro [17, 18]. The cell line was kindly donated by Dieter Gruenert (Cardiovascular Research Institute, University of California, San Francisco).

3. Caco-2: intestine epithelial cell line (Fig. 1c).

   Caco-2 is a human intestinal cell line extracted from a colon adenocarcinoma [19]. The cell line undergoes spontaneous differentiation in culture, leading to monolayer formation and expression of morphological and functional characteristics of mature intestinal enterocytes [20]. Caco-2 cells show a polarized morphology including microvilli formation on the apical side and form tight junctions between neighboring cells [20]. The applications of the cell line include permeability and absorption testing across the intestinal epithelium [19, 21]. The cell line is available from ATCC (HTB-37™, LGC Standards, Germany).

2.3 Cell Culture (See Note 1)

1. Medium for A549 cells: Roswell Park Memorial Institute 1640 Medium (RPMI-1640 medium) supplemented with 10% (v/v) fetal bovine serum (FBS), 1% (v/v) L-glutamine (L-Glu), and 1% (v/v) penicillin/streptomycin (P/S).

2. Medium for 16HBE14o- cells: Minimum Essential Medium (MEM) supplemented with 10% (v/v) FBS, 1% (v/v) L-Glu, and 1% (v/v) P/S.

3. Medium for Caco-2 cells: Dulbecco's Modified Eagle's Medium (DMEM) supplemented with 10% (v/v) FBS, 1% (v/v) L-Glu, 1% (v/v) P/S, and 1% (v/v) MEM nonessential amino acids solution (100X).

4. 1X phosphate-buffered saline (PBS).

5. 0.05% trypsin/EDTA.

6. T-25 and T-75 cell culture flasks (*see* **Note 2**).

2.4 Bioprinting

2.4.1 Bioprinter Platform and Control Software

3D Discovery™ is a 3D bioprinting instrument (3DD-G01, Ref. 900002770, regenHU Ltd., Switzerland), allowing the spatially controlled deposition of cells or biomaterials. It is located within a biosafety cabinet (BSC-3DD, Ref. 900005986, regenHU Ltd., Switzerland) to accommodate printing under sterile conditions (Fig. 2a). The bioprinter is equipped with three inkjet printheads (CF-300N, Ref. 900002773, regenHU Ltd., Switzerland) and one direct dispenser (DD-135N, Ref. 900002772, regenHU Ltd., Switzerland) (Fig. 2b).

The inkjet printheads enable jetting or contact dispensing and support viscosities in the range of 110–1,000 mPa·s. The direct dispenser enables contact dispensing using time–pressure technology in a viscosity range of 50–200,000 mPa·s. The inkjet printhead is equipped with a cartridge cooler (CU-05, Ref. 900003936; CC-03CC, Ref. 900003932, regenHU Ltd., Switzerland) to maintain constant temperatures along the cartridge, allowing printing of thermosensitive materials.

The printing platform comprises a cartridge (C-03CC-T, Ref. 602002748, regenHU Ltd., Switzerland), which is connected to a luer lock adapter (LUER-AD-T, Ref. 602002742, regenHU Ltd., Switzerland) on the lower part. On the upper part, the cartridge is sealed with a cartridge pressure adapter (CAH-03CC, Ref. 602002763, regenHU Ltd., Switzerland), which connects the central air inlet of the bioprinter with the cartridges of each printhead. Pressure, which is required for dispensing the biomaterial from cartridges, is controlled by individual pressure regulators. The luer lock adapter is screwed into the inlet adapter (IA-CF300, Ref. 420011176, regenHU Ltd., Switzerland), whereby the cartridge is fixed with the printhead and connected to the microvalve. For jetting, the microvalve is directly used to dispense biomaterial, while an additional needle is mounted on the microvalve to enable

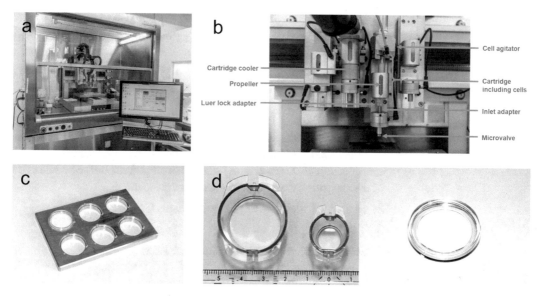

Fig. 2 (a) 3D Discovery™ platform including bioprinting instrument, biosafety cabinet, and human machine interface software. **(b)** Setup of the bioprinter instrument. **(c)** Insert holder containing a Millicell® Cell Culture insert used to avoid insert displacement. **(d)** Falcon® Cell Culture inserts for 6-well plates, 12-well plates, and Millicell® Cell Culture insert for 6-well plates (from left to right)

contact dispensing. The diameter of the microvalve ranges between 0.1, 0.15, or 0.3 mm, while four lengths of needles are available with a diameter of 0.15 and 0.3 mm. In our experiments, we used microvalves with diameters of 0.1 mm to print cells and 0.15 mm for collagen type I printing, respectively (MVJ-D0.1S0.06, Ref. 602002744; MVJ-D0.15S0.1, Ref. 900005895, regenHU Ltd., Switzerland). In addition, a needle with a diameter of 0.3 mm was used to print collagen type I (NCF-D0.3L6, Ref. 900005185, regenHU Ltd., Switzerland) (*see* **Note 3**).

The bioprinter hardware is controlled via the human machine interface (HMI) software (CU-WIN7, Ref. 900006043, regenHU Ltd., Switzerland), which comprises the selection of printhead, position of print table, initiation and termination of the process, and modification of dosing distance (DD) and valve opening time (VOT). The printing pattern is designed in BioCAD (BIOCAD-IL, Ref. 900009433, regenHU Ltd., Switzerland), and printing parameters like microvalve/needle height, feed rate, and well plate size can be adjusted using this software. The generated g-code file of the pattern is loaded to HMI and used after setting the printing parameters as described in Subheading 3. The design of the printing table allows printing both on glass slides and directly in well plates or on inserts. To avoid any displacement of particular cell culture inserts during printing, an insert holder (Fig. 2c) is placed directly on the table. As cells tend to sediment in the cartridge, a propeller (CA-PROP, Ref. 900008086, regenHU Ltd., Switzerland) actuated by a cell agitator (CA, Ref. 900007250, regenHU Ltd., Switzerland) is placed in the cartridge and rotated during the printing process (Fig. 2b).

The bioprinting instrument has a printing resolution of 5 μm and the smallest printable volume is in the nanoliter range.

2.4.2 Additional Materials

1. Inserts (Fig. 2d): Millicell® Cell Culture inserts for 6-well plates (PICM0RG50, Merck Millipore, Switzerland) and Falcon® Cell Culture inserts for 6-well plates (Ref. 353091, BD Biosciences, Switzerland) and 12-well plates (Ref. 353181, BD Biosciences, Switzerland) (*see* **Notes 4** and **5**).

2. 6- and 12-well plates (Falcon Tissue Culture Plates).

3. Tweezers.

4. Sterile purified water or sterile 1X PBS.

5. 70% ethanol.

2.4.3 Printing Cell Substrate on Millicell® Cell Culture Inserts

While printing the cells on Millicell® Cell Culture inserts, an insert coating to allow cell attachment is required. Falcon® Cell Culture inserts do not necessitate any coating (*see* **Note 6**).

Collagen type I is the most abundant protein in mammals [22] and was therefore chosen as a cell substrate. For coating 6 Millicell® Cell Culture inserts, prepare 450 μL collagen type I solution (1.5 mg/mL) by mixing 225 μL rat tail collagen type I (3 mg/mL, A1048301, ThermoFisher Scientific, Switzerland) with 88.5 μL buffer solution and 136.5 μL cell culture medium of the cell line. The buffer solution consists of 30 μL 10X PBS with 7.5 μL 1 M NaOH, 6 μL 7.5% (w/v) NaHCO$_3$, and 45 μL 1 M HEPES. Prepare the solution immediately before printing and store at 0–5 °C (*see* **Note 7**).

2.5 Lactate dehydrogenase (LDH) Assay

To quantify cell viability after printing, a colorimetric assay can be performed. The LDH assay is based on the measurement of LDH activity released from cells, as an indicator of cell membrane damage and therefore cell death [23].

For the performance of the LDH assay, the following materials are required:

1. 96-well plate.

2. 300 μL medium-based supernatant of printed samples.

3. 300 μL medium-based supernatant of negative control samples (i.e., untreated cells).

4. 300 μL medium-based supernatant of positive control samples (with 0.2% Triton™ X-100 treated cells).

5. Cell culture medium and distilled water.

6. Cytotoxicity detection kit (Ref. 11 644 793 001, Sigma-Aldrich, Switzerland): catalyst and dye solution.

7. Centrifuge with rotor for spinning the supernatant.

8. Microplate reader (e.g., Bio-Rad Benchmark plus Microplate Spectrophotometer reader) with 490–492 nm filter for absorbance measurement and a filter over 600 nm for the reference wavelengths.

2.6 Transepithelial Electrical Resistance (TEER)

Epithelial cells express tight junctions that connect individual cells in an epithelial sheet restricting the paracellular permeability [24]. TEER measures the electrical resistance between the two milieus separated by the epithelial sheet. This value resembles the quality of the epithelial sheet formation.

The following are required to measure TEER:

1. Millicell® Electrical Resistance System Voltohmmeter (MERS00002, Merck Millipore, Switzerland).

2. 70% ethanol.

3. 1X PBS.

4. 12-well insert (Ref. 353181, BD Biosciences, Switzerland) for reference measurements.

2.7 Cell Staining for Fluorescence Confocal Laser Scanning Microscopy

1. 4% (w/v) paraformaldehyde (PFA, Sigma-Aldrich, Switzerland) in 1X PBS: add PFA to 1X PBS and stir it at 125 °C for 2 h until the powder has dissolved. Let the solution cool down and store aliquots at −20 °C.

2. 0.1 M glycine (Ref. 15527013, ThermoFisher Scientific, Switzerland) in 1X PBS.

3. 0.2% (v/v) Triton™ X-100 (Ref. T8787, Sigma-Aldrich, Switzerland) in 1% (w/v) bovine serum albumin (BSA, Sigma-Aldrich, Switzerland) in 1X PBS: for 100 mL, add 200 μL Triton™ X-100 to 1% BSA in 1X PBS.

4. Washing solution: 1X PBS and 0.1% (w/v) BSA in 1X PBS.

5. Fluorescent dyes: 4′,6-diamidino-2-phenylindole (DAPI, D1306, ThermoFisher Scientific, Switzerland) and rhodamine-phalloidin (R415, ThermoFisher, Switzerland).

6. Glycergel® (Dako, Denmark).

7. Microscopy glass slides and 0.17 mm thick glass coverslips.

3   Methods

3.1 Cell Culture

3.1.1 Alveolar Epithelial Cell Line A549

To seed A549 cells, first add 8–10 mL complete RPMI medium to a 25 cm² cell culture flask. Next, put the flask in the incubator to allow the medium to pre-equilibrate to temperature and pH prior to cell thawing. Remove the vial containing the frozen cells from the nitrogen tank, and warm it up in the water bath at 37 °C. Avoid immersion of the lid into the water bath to prevent contamination. When the ice has melted, sterilize the vial by wiping it with 70% ethanol and place it in the sterile cell culture hood. Pipette the cells in the prepared flask and rock the flask gently to distribute the cells within the flask. Let the cells adhere for 12 h. Next, aspirate the medium and replace it with fresh, prewarmed medium. Culture the cells until they reach 70–80% confluency and continue to subculture.

When the cells have reached 70–80% confluency, aspirate the medium and wash the cells gently with 1X PBS. Add 2 mL trypsin/EDTA (for T-75 flask) to the flask and cover the monolayer completely. Put the flask in the incubator and check if the cells have detached after 3 min. As soon as all cells are detached, add medium to neutralize trypsin/EDTA, and pipette up and down to disperse cell clusters. Dilute the cells 1:10–1:16 and allow the cells to culture until they reach 70–80% confluency. Use cells for experiments starting from passage 3 after thawing until passage number 20–25.

3.1.2 Bronchial Epithelial Cell Line 16HBE14o-

16HBE14o- cells are cultured similarly to A549 cells but with minor modifications. Prepare a cell culture flask for sensitive adherent cells, and add 8–10 mL complete MEM. Dilute the cells 1:10 once a week and change medium every 3–4 days.

3.1.3 Intestine Epithelial Cell Line Caco-2

For seeding Caco-2 cells, prepare a 25 cm^2 flask and add 8–10 mL complete DMEM. Seed cells into flask and subculture according to the A549 protocol described above. Use a maximum splitting dilution of 1:4, and subculture twice a week. Change medium every 2 days.

3.2 Bioprinting Parameters

For optimizing bioprinting, the 3D Discovery platform provides several adjustable parameters. The HMI software controls the parameters DD and VOT, influencing the dispensed volume of cells or cell substrates. The DD defines the distance between two dispensed droplets, resulting in printing that varies from dots to a continuous line. The VOT specifies how long the microvalve is opened to dispense material and defines the volume of a printed droplet, influencing the resolution of the printed pattern. To obtain highly resolved printing patterns, the shortest possible VOT should be chosen. In our lab, it was observed that 120 μs is the shortest period to dispense cells or cell substrate. Using BioCAD software, the printing pattern, the feed rate, and the microvalve/needle height are adjusted. To design the printing pattern, different drawing options (circle, line, dots, and text) can be chosen, which are adapted to the size of the insert. The feed rate specifies the speed of the printhead during the printing process, and the microvalve/needle height describes the initial distance between the printing table and the microvalve/needle. Once these parameters have been

Table 1
Summary of the parameters used for bioprinting epithelial cells using the 3D Discovery™ platform

	Printing collagen type I solution	Printing cells
Printing temperature	5 °C	Room temperature
Microvalve diameter (jetting technique)	0.15 mm	0.1 mm
Needle diameter (contact dispensing technique)	0.3 mm	–
Feed rate	20 mm/s	20 mm/s
Pressure (*see* **Note 8**)	0.25 bar	0.25 bar
Dosing distance	0.3–0.5 mm	0.1–0.7 mm
Valve opening time	120 μs	120 μs

set, a g-code file is generated and will be loaded in HMI software. The pressure is controlled directly on the 3D Discovery instrument by manipulating the individual pressure regulator connected to the printheads.

3.2.1 Parameters for Bioprinting Epithelial Cell Layers

Table 1 shows the summary of parameters used for bioprinting epithelial cell layers:

The microvalve and needle height differ between the two used inserts. For jetting, the microvalve height is 11 mm for Millicell® Cell Culture inserts and 19 mm for Falcon® Cell Culture inserts, while for contact dispensing, the needle height is 6 and 0.5 mm, respectively.

Cell substrate is printed on Millicell® Cell Culture inserts in circles to obtain a homogenous coating. As the diameter of microvalve and needle are different, line spaces of 0.1 and 0.3 mm are used for jetting and contact dispensing, respectively. We decided to print cells in two different configurations: (a) complete coverage of the inserts (Millicell® Cell Culture inserts) and (b) partial coverage allowing the cells to grow and migrate (Falcon® Cell Culture inserts). For complete coverage, cells are printed in circles with a line space of 0.25 mm in the middle and 0.2 mm at the edges to counteract the tendency of cells in suspension to accumulate in the center of the insert. Partial coverage is achieved by printing lines with a spacing of 1 mm for 6-well and 0.5 mm for 12-well inserts. The diameter of the printing pattern is 22 mm for 6-well inserts and 10 mm for 12-well inserts.

3.3 Bioprinting

Prepare the bioprinting platform by starting the laminar flow, then wipe the surfaces with 70% ethanol followed by a 30 min UV light irradiation. Use only autoclaved cartridges and luer lock adapters and sonicate the microvalve, needle, inlet adapter, and propeller in absolute ethanol for 20–30 min. Perform the following printing steps only under sterile and safe conditions:

Remove the sonicated microvalve, needle, inlet adapter, and propeller from absolute ethanol, and let it air dry inside the hood under the laminar flow. Mount the bioprinting equipment according to the manufacturer's instructions using sterile tweezers. Measure the length of the microvalve/needle to allow an exact bioprinting process using HMI. Add 1 mL 70% ethanol to the cartridge, and close it with the cartridge pressure adapter. Apply pressure and dispense 250–500 droplets to test the microvalve/needle. If successful, rinse the system and repeat it with 1 mL sterile purified water (*see* **Note 9**).

3.3.1 Printing Cell Substrate for Millicell® Cell Culture Inserts

Remove the cartridge pressure adapter from the cartridge, and add the collagen type I solution to the precooled (5 °C) cartridge using a pipette (*see* **Note 7**). Set the printing parameters as summarized in Subheading 3.2.1, and test the printer by dispensing 250–500 droplets. Load the g-code file of the printing pattern, place the inserts in the insert holder, and start bioprinting. After finishing bioprinting, remove the inserts from the holder using tweezers and place them in a well plate to ensure sterile conditions. Add 1 mL 1X PBS to the bottom of the inserts, and let the inserts in the incubator for at least 30 min to allow collagen type I to polymerize. Afterward, add 1 mL sterile 1X PBS to the top of the insert to prevent drying. The coated inserts can be stored in 1X PBS in the incubator at 37 °C.

3.3.2 Printing Cells

Only use the cells when they have reached the minimum 80% confluency. Detach the cells and determine the cell concentration using a cell counter. Spin the cells down, aspirate the medium, and add cell culture medium. The cell concentration used for bioprinting depends on the cell line and has to be optimized for each application. We used concentrations between 3 and 15 million cells/mL and obtained satisfactory results when using 5 million cells/mL for A549 and 10 million cells/mL for 16HBE14o- and Caco-2 cells (*see* **Note 10**).

Remove the cartridge pressure adapter from the cartridge, and add the cells to the cartridge using a pipette. Use the propeller to agitate the cells. Set the printing parameters as summarized in Subheading 3.2.1, and test the printer by dispensing 250–500 droplets. Load the g-code file of the printing pattern, place the Millicell® Cell Culture inserts in the insert holder or the Falcon® inserts in the well plate (*see* **Note 11**), and start bioprinting. When printing on Falcon® inserts, add 1.2 mL (6-well) or 600 μL (12-well) of medium to the bottom of the wells (below the inserts), to prevent drying of the recently printed cells. When Millicell Cell Culture inserts are used, place them after the printing process in a 6-well plate containing 1.5 mL medium. Let the cells attach (1 h for

A549, 1 h for 16HBE14o-, and 5 h for Caco-2 cells), and then add 1–2 mL of medium on the top.

3.4 LDH Assay

1. Prepare the positive control by adding Triton™ X-100 in 1X PBS to the sample to obtain a final concentration of 0.2% Triton™ X-100. Keep it for 30 min in the incubator.

2. Collect the supernatant of the treated samples and of the negative and positive controls and centrifuge for 5 min at $250 \times g$.

3. Use a 96-well plate and prepare triplicates with 100 µL supernatant of the treated samples and the negative control. Dilute the supernatant of the positive control 1:5 or 1:10 to avoid too high absorbance values and prepare triplicates as well. Prepare triplicates of medium and distilled water as control.

4. Prepare the working solution of the cytotoxicity detection kit by mixing 2.5 µL catalyst with 112.5 µL dye solution per well, and add 100 µL to each sample using a multichannel pipette.

5. Measure the absorbance at 490 nm, and use a reference wavelength of 630 nm. The reference wavelength is used to normalize influences of the cytotoxicity detection kit. Either the slope of the absorbance over time or the absolute value can be measured and used for viability testing.

6. The cytotoxicity expressed in percentage (Fig. 3a) can be calculated according to the manufacturer's protocol. Subtract the absorbance value of the background control (e.g., cell culture medium) from the average absorbance value of the triplicates. Use the equation below to calculate the percentage cytotoxicity:

$$\text{Cytotoxicity } [\%] = \frac{\text{exp .value} - \text{low control}}{\text{high control} - \text{low control}} \times 100$$

3.5 TEER

The TEER of the samples is measured using a Millicell® Electrical Resistance System Voltohmmeter. To obtain the tissue resistance, the TEER of a blank insert should be measured additionally. Please note that TEER measurement using this method only works accurately for 12-well inserts or smaller, as the manufacturer does not guarantee a continuous distribution of the current across 6-well inserts and readings are overestimated (*see* **Note 12**).

3.5.1 TEER Measurement

1. Perform the following steps in a laminar flow hood and under sterile conditions.

2. Sterilize the electrode by immersing it in 70% ethanol for 15 min. Let it air dry for 15 s and rinse in 1X PBS.

3. Remove the medium from the bottom and top of the inserts, and wash the samples once with 1X PBS. Add 1.5 mL 1X PBS

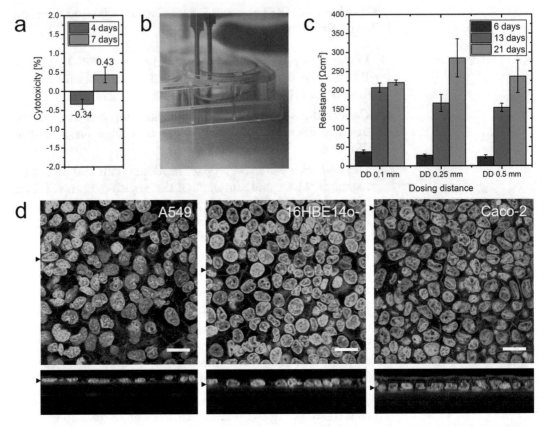

Fig. 3 (**a**) Percentage cytotoxicity of printed A549 after 4 and 7 days. (**b**) Handling of the electrode during TEER measurements. In (**c**) TEER measurements of Caco-2 cells printed with different dosing distances (0.1, 0.25, and 0.5 mm) after 6, 13, and 21 days are shown (triplicate measurements were performed per time point). (**d**) Confocal fluorescence laser scanning microscope images of the cell lines A549, 16HBE14o-, and Caco-2 using DAPI and rhodamine-phalloidin staining of cell nuclei (cyan) and F-actin (magenta), respectively. The cells were printed in Falcon® Cell Culture inserts and cultivated for 7 (A549, 16HBE14o-) and 21 days (Caco-2). Scale bar = 20 μm

below and 1 mL above the inserts, so that the electrode is fully immersed in 1X PBS (12-well inserts). Place the electrode at a 90° angle to the insert with the shorter tip in the insert and the longer tip in the outer well (Fig. 3b). Avoid any contact with growing cells. The measured resistance can be read from the voltohmmeter.

4. Perform the measurements in triplicate.

3.5.2 Resistance Calculations

1. To obtain the tissue resistance, the resistance across the blank insert is subtracted from the resistance measured across the inserts seeded with cells.

2. For consistency across different well plate sizes, the resistance is multiplied by the area of the membrane area for a value that is independent of the area of the used membrane (Fig. 3c).

3.6 Cell Staining with Fluorescence Laser Scanning Microscopy for Image Acquisition

3.6.1 Cell Staining

1. Remove the cell culture medium, and wash the cells once with 1X PBS. Fix cells within a fume hood in 4% PFA for up to 15 min, and wash the samples three times in 1X PBS. If the staining cannot be performed on the same day, store the samples in 1X PBS at 0–5 °C. Do not let the samples dry out, and seal the well plate using Parafilm® (Sigma-Aldrich) (*see* **Note 13**).

2. Aspirate PBS and add 0.1 M Glycine for 15 min to remove residual PFA. Wash the cells three times with 1X PBS.

3. To permeabilize the cell membrane, add 0.2% Triton™ X-100 in 1% BSA in 1X PBS and immerse the samples for 15 min. Afterward, wash the cells three times with 0.1% BSA in 1X PBS.

4. To label the F-actin cytoskeleton, dilute rhodamine-phalloidin 1:200 in 0.1% BSA in 1X PBS, and add it to the cells, so that the whole sample is covered. Immerse the sample for 60 min in the dark, and wash three times with 0.1% BSA in 1X PBS. To stain the cell nuclei, dilute DAPI 1:100 in 0.1% BSA in 1X PBS and add it to the cells for 5 min. Keep the samples in the dark during the incubation. Wash the cells three times with 1X PBS.

5. Mount the samples on glass slides by adding 50 μL Glycergel on the bottom and top and cover it with a coverslip. Store the samples in the dark at 0–5 °C.

3.6.2 Image Acquisition

The image acquisition can be performed using an inverted confocal laser scanning microscope using the excitation lasers at 405 nm for DAPI and 561 nm for rhodamine-phalloidin. The emission of the fluorophores is collected in their respective emission wavelengths. The acquired images are processed using Fiji and Imaris (Bitplane AG) and converted to RGB format. Cell structures are displayed in cyan (nuclei) and magenta (F-actin), and orthogonal views are prepared from stacked acquisitions (Fig. 3d).

4 Notes

1. The complete media must be stored at 0–5 °C and used within 30 days or aliquoted and stored for long-term use at −20 °C.

2. For subculturing 16HBE14o- cells, cell culture flasks for sensitive adherent cells facilitating cell attachment are necessary.

3. When printing a defined pattern or using more viscous materials, contact dispensing may be better than jetting. Needles with

a length of 2.4, 6, 22, and 26 mm and a diameter of 0.15 mm or 0.3 mm are available and can be used for these purposes.

4. 16HBE14o- cells grow better with a fibronectin/collagen coating to allow cell attachment when printed on cell culture inserts.

   1 mL fibronectin/collagen coating solution: add 10 μL human fibronectin (1 mg/mL, BD Laboratories), 10 μL bovine collagen I (3 mg/mL, Vitrogen bovine collagen, Cohesion Technologies, FXP-019), and 100 μL BSA in basal medium (1 mg/mL, Sigma-Aldrich) to 880 μL 1X PBS.

   Add the prepared coating solution to the inserts and ensure that the bottom is covered, i.e., 400 μL for a 6-well insert or 200 μL for a 12-well insert. Incubate for 3 h at 37 °C or overnight at room temperature. Thereafter, remove the liquid and store the inserts at 4 °C (up to 1 month).

5. Printing small dots or thin lines (e.g., a few mm) directly in wells (without inserts) cannot be performed. The cell culture medium evaporates before the cells can attach leading to reduced viability.

6. Caco-2 cells are very sensitive cells and need to be seeded at a high density to form a continuous monolayer and display a long attachment time. Printing Caco-2 cells on coated Falcon Cell Culture inserts according to **Note 4** may improve confluent cell layer formation.

7. Polymerization of collagen type I is initiated by increasing the pH and temperature of the solution to physiological conditions [25, 26]. To avoid any polymerization during preparation of the collagen type I solution, work quickly and keep the solution in the fridge. Additionally, ensure that the cartridge is precooled to 5 °C before adding the collagen type I solution.

8. The pressure used during the printing process can be varied to modify the volume of the printed material. We have observed that pressure up to 2 bar does not reduce cell viability, however the high pressure causes splashes, which may lead to cross-contamination. Additionally, the printing pattern is more precise when using a pressure around 0.25 bar.

9. Instead of sterile purified water, sterile 1X PBS can be used, being pre-equilibrating the pH of the printing equipment. However, due to the composition of 1X PBS, clogging of the needle may occur.

10. When using a valve with a diameter of 0.1 mm, up to 15 million cells/mL may be printed without clogging the valve. Less than 1 million cells/mL are deemed insufficient.

11. Falcon Cell Culture inserts are usually not fixed in wells and tend to displace during handling of the well plate. While

printing, both the printhead and the print table with the well plate move, leading to a displacement of the inserts within the wells. This displacement affects the pattern and should be avoided. Inserts can be fixed within the wells using a rubber band or a purpose-built lid. In our lab, it was observed that the inserts displaced always to the right lower side of the well. Therefore, the inserts were moved to the right lower part of the well and the printing range was adjusted. To be sure that the inserts are at the correct position, the printhead moves first inside the inserts and moves them to the right location. After initiation of the printing process, the valve remains above the inserts.

12. To receive more reliable data, measure TEER for each insert three times and on both sides. Ensure that the electrode is placed vertically and does not touch the insert. Touching of the insert could scrape cells and affect the outcome. The electrode should be rinsed in PBS from time to time to ensure stable measurements.

13. When preparing the inserts for staining, only use half of the insert and store the other half in 1X PBS in the fridge as backup.

5   Conclusions

Much effort has been made over the last several years for systematic development and evaluation of innovative and more reliable epithelial tissue models in the hope of improving R&D productivity in the pharmaceutical and biomedical fields. However, fully validated and robust cell and tissue culture solutions are yet to be achieved. The use of 3D printing technology for printing biological materials, i.e., biomaterials as well as cells, in itself is an innovative and new approach. While initial results are very promising, the control and standardization of the process for reproducibility of printed samples require significant improvement. To this end, we have identified optimal parameters for printing epithelial cell layers as the basis for more complex respiratory and gastrointestinal tissue models. Such systems will be useful for high-throughput toxicity screening and drug efficacy evaluation.

Acknowledgments

This work was supported by the Swiss National Science Foundation (grant number CRSII5_171037), the Run4Science grant, and the Adolphe Merkle Foundation.

References

1. Stern ST, McNeil SE (2008) Nanotechnology safety concerns revisited. Toxicol Sci 101:4–21. https://doi.org/10.1093/toxsci/kfm169

2. Hartung T, Rovida C (2009) Chemical regulators have overreached. Nature 460:1080–1081. https://doi.org/10.1038/4601080a

3. Shukla SJ, Huang R, Austin CP, Xia M (2010) Foundation review: the future of toxicity testing: a focus on in vitro methods using a quantitative high-throughput screening platform. Drug Discov Today 15:997–1007. https://doi.org/10.1016/j.drudis.2010.07.007

4. Astashkina A, Mann B, Grainger DW (2012) A critical evaluation of in vitro cell culture models for high-throughput drug screening and toxicity. Pharmacol Ther 134:82–106. https://doi.org/10.1016/j.pharmthera.2012.01.001

5. Xia M, Huang R, Witt KL et al (2008) Compound cytotoxicity profiling using quantitative high-throughput screening. Environ Health Perspect 116:284–291. https://doi.org/10.1289/ehp.10727

6. OECD (2015) Test No. 439: In vitro skin irritation: reconstructed human epidermis test method. OECD Publishing, Paris

7. Murphy SV, Atala A (2014) 3D bioprinting of tissues and organs. Nat Biotechnol 32:773–785. https://doi.org/10.1038/nbt.2958

8. Wüst S, Godla ME, Müller R, Hofmann S (2014) Tunable hydrogel composite with two-step processing in combination with innovative hardware upgrade for cell-based three-dimensional bioprinting. Acta Biomater 10:630–640. https://doi.org/10.1016/J.ACTBIO.2013.10.016

9. Pati F, Jang J, Ha D-H et al (2014) Printing three-dimensional tissue analogues with decellularized extracellular matrix bioink. Nat Commun 5:3935. https://doi.org/10.1038/ncomms4935

10. Horváth L, Umehara Y, Jud C et al (2015) Engineering an in vitro air-blood barrier by 3D bioprinting. Sci Rep 5:7974. https://doi.org/10.1038/srep07974

11. Lieber M, Todaro G, Smith B et al (1976) A continuous tumor-cell line from a human lung carcinoma with properties of type II alveolar epithelial cells. Int J Cancer 17:62–70. https://doi.org/10.1002/ijc.2910170110

12. Shapiro DL, Nardone LL, Rooney SA et al (1978) Phospholipid biosynthesis and secretion by a cell line (A549) which resembles type II alveolar epithelial cells. Biochim Biophys Acta 530:197–207. https://doi.org/10.1016/0005-2760(78)90005-X

13. Foster KA, Oster CG, Mayer MM et al (1998) Characterization of the A549 cell line as a type II pulmonary epithelial cell model for drug metabolism. Exp Cell Res 243:359–366. https://doi.org/10.1006/EXCR.1998.4172

14. Foldbjerg R, Dang DA, Autrup H (2011) Cytotoxicity and genotoxicity of silver nanoparticles in the human lung cancer cell line, A549. Arch Toxicol 85:743–750. https://doi.org/10.1007/s00204-010-0545-5

15. Forbes B, Ehrhardt C (2005) Human respiratory epithelial cell culture for drug delivery applications. Eur J Pharm Biopharm 60:193–205. https://doi.org/10.1016/J.EJPB.2005.02.010

16. Lehmann A, Brandenberger C, Blank F et al (2010) A 3D model of the human epithelial airway barrier. In: Maguire D, Novik E (eds) Methods in bioengineering: Alternative technologies to animal testing. Artech House, London, pp S35–S36

17. Wan H, Winton HL, Soeller C, et al (2000) Tight junction properties of the immortalized human bronchial epithelial cell lines Calu-3 and 16HBE14o-. Eur Respir J 15:1058–1068

18. Cozens AL, Yezzi MJ, Kunzelmann K et al (1994) CFTR expression and chloride secretion in polarized immortal human bronchial epithelial cells. Am J Respir Cell Mol Biol 10:38–47. https://doi.org/10.1165/ajrcmb.10.1.7507342

19. Hidalgo IJ, Raub TJ, Borchardt RT (1989) Characterization of the human colon carcinoma cell line (Caco-2) as a model system for intestinal epithelial permeability. Gastroenterology 96:736–749

20. Sambuy Y, De Angelis I, Ranaldi G et al (2005) The Caco-2 cell line as a model of the intestinal barrier: influence of cell and culture-related factors on Caco-2 cell functional characteristics. Cell Biol Toxicol 21:1–26. https://doi.org/10.1007/s10565-005-0085-6

21. Artursson P, Karlsson J (1991) Correlation between oral drug absorption in humans and apparent drug permeability coefficients in human intestinal epithelial (Caco-2) cells. Biochem Biophys Res Commun 175:880–885. https://doi.org/10.1016/0006-291X(91)91647-U

22. Fratzl P (2008) Collagen: structure and mechanics, an introduction. In: Collagen. Springer US, Boston, MA, pp 1–13

23. Legrand C, Bour JM, Jacob C et al (1992) Lactate dehydrogenase (LDH) activity of the number of dead cells in the medium of cultured eukaryotic cells as marker. J Biotechnol 25:231–243. https://doi.org/10.1016/0168-1656(92)90158-6

24. Balda MS, Whitney JA, Flores C et al (1996) Functional dissociation of paracellular permeability and transepithelial electrical resistance and disruption of the apical-basolateral intramembrane diffusion barrier by expression of a mutant tight junction membrane protein. J Cell Biol 134:1031–1049. https://doi.org/10.1083/JCB.134.4.1031

25. Velegol D, Lanni F (2001) Cell traction forces on soft biomaterials. I. Microrheology of type I collagen gels. Biophys J 81:1786–1792. https://doi.org/10.1016/S0006-3495(01)75829-8

26. Williams BR, Gelman RA, Poppke DC, Piez KA (1978) Collagen fibril formation. Optimal in vitro conditions and preliminary kinetic results. J Biol Chem 253:6578–6585

Chapter 14

Bioprinting for Skin

Cristina Quílez, Gonzalo de Aranda Izuzquiza, Marta García,
Verónica López, Andrés Montero, Leticia Valencia, and Diego Velasco

Abstract

We describe an extrusion-based method to print a human bilayered skin using bioinks containing human plasma and primary human fibroblasts and keratinocytes from skin biopsies. We generate 100 cm^2 of printed skin in less than 35 min. We analyze its structure using histological and immunohistochemical methods, both in in vitro 3D cultures and upon transplantation to immunodeficient mice. We have demonstrated that the printed skin is similar to normal human skin and indistinguishable from bilayered dermo-epidermal equivalents, previously produced manually in our laboratory and successfully used in the clinic.

Key words 3D bioprinting, Skin bioprinting, Skin tissue engineering, Artificial skin, Skin equivalents, 3D skin culture, Fibrin hydrogel

1 Introduction

Based on the 3D printing technologies and on the concepts developed in Tissue Engineering during the last decades, 3D bioprinting is emerging as one of the most innovative and promising technologies for the generation of human tissues and organs [1–5]. Regarding skin bioprinting, using different bioprinting methods and bioinks, interfollicular skin has been recently generated with a structural and functional quality that paves the way for clinical and industrial applications [6–11]. We have acquired extensive expertise in the manual production of autologous human bilayered skin equivalents to treat burns and traumatic and surgical wounds in a large number of patients in Spain and for the generation of skin-humanized mouse models [12–17]. Recently, our laboratories have developed a complete system (continuous extrusion bioprinter and bioinks) to produce bilayered human skin. The lower layer of the generated skin is composed of a human plasma-

Cristina Quílez and Gonzalo de Aranda Izuzquiza contributed equally to this work.

Jeremy M. Crook (ed.), *3D Bioprinting: Principles and Protocols*, Methods in Molecular Biology, vol. 2140,
https://doi.org/10.1007/978-1-0716-0520-2_14, © Springer Science+Business Media, LLC, part of Springer Nature 2020

derived fibrin matrix populated with primary human fibroblasts (hFbs). The upper layer consists of confluent primary human keratinocytes (hKcs), deposited on top of the fibrin scaffold [18]. We analyze the structure and function of the printed skin using histological and immunohistochemical methods, both in 3D in vitro cultures and after long-term transplantation to immunodeficient athymic mice (skin-humanized mice). For the in vitro assays, skin constructs deposited on transwells are allowed to differentiate at the air–liquid interface for 17 days at 37 °C in a CO_2 incubator in a differentiation medium. For the in vivo assays, once hKcs have attached to the fibrin matrix, the printed equivalents are manually detached from the culture plate and grafted onto the backs of immunodeficient mice. The grafts are analyzed 8 weeks after grafting. Our results show that the printed skin has structural and functional characteristics similar to those of normal human skin and of skin equivalents produced manually by our group. We also demonstrate the capacity of our process to reproducibly print large areas of human skin, useful for the treatment of diverse cutaneous pathologies such as burns, ulcers, and surgical wounds.

2   Materials

2.1 Cells

Primary human fibroblasts, hFbs, and human keratinocytes, hKcs, are obtained from skin biopsies of healthy donors, amplified in in vitro culture and deposited and registered in the collection of biological samples of human origin in the "Registro Nacional de Biobancos para Investigación Biomédica del Instituto de Salud Carlos III." Confluent lethally irradiated 3T3-J2, or Swiss albino mouse 3T3 fibroblasts are used as feeder layers to culture hKcs.

2.2 Growth Media

(a) Feeder culture medium (FCM).
   - 3:1 mixture of Dulbecco's Modified Eagle Medium (DMEM) and HAM'S F12.
   - 0.1 nM choleric toxin.
   - 2 nM triiodothyronine.
   - 5 mg/ml insulin.
   - 0.4 mg/ml hydrocortisone.
   - 24 µg/ml adenine.
   - +10% v/v of fetal bovine serum (FBS).

(b) Keratinocyte culture medium (KCM).
   - 3:1 mixture of DMEM and HAM'S F12.
   - 0.1 nM choleric toxin.
   - 2 nM triiodothyronine.
   - 5 mg/ml insulin.

- 0.4 mg/ml hydrocortisone.

- 24 µg/ml adenine.

- 10 ng/ml epidermal growth factor (EGF).

- +10% v/v of fetal bovine serum (FBS).

(c) Fibroblast culture medium (FbCM).
  - DEMEM.

  - +10% w/v FBS.

(d) Differentiation culture medium (DCM).
  - KC culture medium.

  - +0.5% w/v FBS.

2.3 Solutions

0.9% NaCl (0.9% w/v) sodium chloride in phosphate-buffered saline (PBS).

0.1% $CaCl_2$ (1% w/v) calcium chloride in 0.9% NaCl.

Fresh frozen human plasma was provided by a local blood bank (Banco de Sangre del Centro Comunitario de Transfusión del Principado de Asturias (CCST), Spain) and was obtained and stored according to the standards of the American Association of Blood Banks [19].

2.4 Bioinks

(a) Bioink 1.
  - hFbs 5×10^5 cells/ml in FbC.

  - 20% v/v tranexamic acid in water (Amchafibrin, Meda Pharma).

(b) Bioink 2.
  - Human plasma containing fibrinogen, diluted with 0.9% NaCl to obtain a final fibrinogen concentration of 1.32 mg/ml.

(c) Bioink 3.
  - 0.1% $CaCl_2$.

(d) Bioink 4.
  - hKcs: 5×10^5 cells/ml in KCM.

2.5 Immunodeficient Mice

- Immunodeficient athymic nude mice (nu/nu) were purchased from IFFA-Credo-Charles River.

- The animals were housed in individually ventilated type II cages containing a maximum of four mice per cage, with 25 air changes per hour and 10 KGy gamma irradiated soft wood pellets as bedding.

2.6 Bioprinter

A multisyringe 3D bioprinter (Fig. 1) was designed based on a regular CNC 3D printer (*see* **Note 1**). Structural parts were

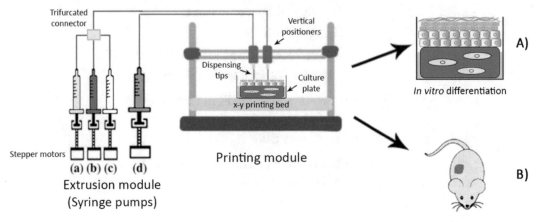

Fig. 1 Scheme of the bioprinting process. The extrusion module contains four syringes, loaded with hFbs (a, yellow, bioink 1), plasma (b, red, bioink 2), CaCl₂ (c, white, bioink 3), and hKc (d, orange, bioink 4), respectively. The contents of the syringes a, b, and c are continuously pumped out at the appropriate volumes and mixed in the trifurcated connector. The mixture is extruded through the dispensing tips and deposited on the corresponding culture plate type (e.g., P100 or transwell), following the trajectories dictated by the computer software. This mixture is allowed to polymerize for 30 min at 37 °C to form a fibroblast-containing fibrin hydrogel, constituting the dermal compartment of the skin equivalent. Afterward, the hKcs suspension contained in syringe (d) is similarly deposited on top of this hydrogel to form a confluent monolayer. (**A**) Equivalents printed in transwell inserts were allowed to differentiate at the air–liquid surface for 17 days and then analyzed. (**B**) Equivalents printed on cell culture plates were grafted onto the backs of immunodeficient mice for 8 weeks and then analyzed. (Modified with permission from [18])

generated by a normal 3D printer using acrylonitrile-butadiene-styrene (ABS). The bioprinter is composed of the following modules:

- Printing module:
  - Two stepper motors control the position of a printing bed in the *x–y* plane. The printing bed is a heated surface in order to maintain the temperature at 37 °C.
  - Two vertical positioners control the distance of deposition between the dispensing tip and the printing bed. A stepper motor moves each of the positioners through the *z*-axis.

- Extrusion module:
  - Four syringe pumps, each one independently driven by one stepper motor, control four sterile plastic syringes containing the bioinks.
  - Three syringes containing bioink 1, 2, and 3 converge into a trifurcated connector where their contents are mixed (*see* **Note 2**).
  - The mixture goes through a dispensing tip (diameter: 0.4 mm) located at a vertical positioner.

- The union between the syringes, the trifurcated connector, and the dispensing tip is made by luer to barb connectors and a sterile medical grade silicone tubing (*see* **Note 3**).

- The flow of each syringe is set to obtain a homogeneous mixture in the dispensing tip with the desired bioink component proportion. The flow in the dispensing tip is 12 ml/min.

- The fourth syringe (bioink 4) is connected to an independent dispensing tip placed in the second vertical positioner at a flow of 4 ml/min.

- Software:
  - Each of the printer modules firmware is installed in the microcontroller (ATmega2560) of a RepRap Arduino Mega Pololu Shield (RAMPS) and manages the mechanical sensors and actuators, as well as the thermal control of the heated bed.

  - The printing module firmware controls the x–y coordinates of the printing bed and the z coordinates of the vertical positioners. Printing trajectories are defined to print on different cell culture plates. The trajectories are executed by the stepper motors that control the x–y plane. Distances in the deposition are defined for the different printing layers. The distances in the deposition are executed by the stepper motors that control the vertical positioners.

  - The extrusion module firmware controls the printed volume and flow rate of the syringe pumps. Each syringe pump is controlled independently.

  - The user interface is executed in the host PC and communicates with the microcontrollers via USB.

2.7 Histology and Immunohisto-chemistry

- 3.7% buffered formaldehyde.
- Tissue-terk (4583, Sakura Finetek, CA).
- Hematoxylin-Eosin (H/E).
- Primary antibodies:
  - Antihuman-vimentin (V9, Bio-Genex, San Ramon).
  - Antikeratin 5 (K5) (polyclonal AF138, BabCO, Berkeley, CA).
  - Antikeratin 10 (K10) (monoclonal AE2, ICN Biomedicals, Cleveland, OH).
  - Antihuman filaggrin (polyclonal AF-62, BabCO).
  - Antihuman-collagen VII (Col VII) (Clone LH7.2, Sigma).
  - Antismooth muscle actin (SMA) (C6198, Sigma, St. Louis, USA).

- Secondary antibodies:
  - Antimouse Alexa 488 (Goat, Thermofisher).
  - Antimouse Alexa 594 (Goat, Thermofisher).
  - Mowiol (Hoechst, Somerville, NJ).
  - Mounting media containing 46-diamidino-2-phenyl indole (DAPI, ROCHE, Germany), 20 μg/ml.

3 Methods

3.1 Fibroblast Culture

1. Human fibroblasts are cultured to confluence in 75 cm^2 culture flasks using FbCM. Cultures are performed at 37 °C and 5% CO_2 in a cell culture incubator.

2. When fibroblasts reach confluency, they are treated with 0.04% trypsin/0.03% EDTA solution (Promocell) to obtain individual cells. After centrifugation at 208 × g for 7 min, they are resuspended in FbCM at the required cell density for bioink 1.

3.2 Keratinocyte Culture

1. Feeder cells (2 × 106 3T3-J2 cells or 5 × 106 Swiss albino mouse 3T3 cells in FCM and irradiated at 50 Gy) are seeded in 75 cm^2 culture flasks and kept in FCM at 37 °C and 5% CO_2 in a cell culture incubator.

2. Add keratinocytes resuspended in FCM in the culture flask without removing the supernatant. Cells are cultured at 37 °C and 5% CO_2 in a cell culture incubator.

3. The medium is changed every 3 days. At the end of the first change, KCM will substitute FCM. From this moment, this will be the culture medium used (*see* **Note 4**).

4. When keratinocyte primary cultures are at approximately 80% confluency (*see* **Note 5**), they are treated with 0.04% trypsin/0.03% EDTA solution (Promocell) for 10 min to obtain individual cells (*see* **Note 6**). After centrifugation, cells are resuspended in KCM culture medium at the required cell density for bioink 4.

3.3 Preparation and Printing

3.3.1 Preparation

1. The printer and the extrusion module are placed in a cell culture laminar flow hood; all the parts are sterilized with UV light for 15 min in the hood.

2. The four bioinks (*see* **Note 7**) are introduced in the four sterile disposable plastic syringes (5 or 20 ml) (*see* **Note 8** and Fig. 1).

3.3.2 Printing the Dermis

1. The syringes containing bioinks 1, 2, and 3 are placed in the extrusion module of the bioprinter and their content is conveniently pumped out of the syringes and mixed at the trifucarted

connector. The volume ratio of the mixture is 5% bioink 1, 87% bioink 2, and 8% bioink 3 (*see* **Note 9**). The final concentrations in the mixture are the following:

- hFbs: 20.000 cell/ml.
- Fibrinogen: 1.2 mg/ml.
- $CaCl_2$: 8% v/v.
- Amchafibrin: 0.8% v/v.

2. The mixture is deposited on the selected culture plate. For in vitro testing, dermal substitutes are printed on polycarbonate transwell inserts (1 μm membrane pore) in a 6-well culture plate (*see* **Note 10**).

3. Printed dermal substitutes are left in a cell culture incubator (at 37 °C in 5% CO2) for 30 min to allow the fibrinogen from human plasma to polymerize into a fibrin hydrogel containing the fibroblasts (dermal equivalent).

3.3.3 Printing the Epidermis

1. The syringe containing bioink 4 is loaded in the extrusion module. The bioink is deposited on top of the dermal equivalent to achieve a seeding density of 8.8×10^5 hKcs/cm^2 (176 μl$_{\text{bioink 4}}$/m^2).

2. For in vivo differentiation (transplant to nu/nu mice), hKcs are allowed to attach and spread overnight in a cell culture incubator. The number of hKcs is established in order to generate a confluent monolayer at this moment.

3. For in vitro test: hKcs are allowed to attach and spread for 48 h in a cell culture incubator.

3.4 In Vivo Maturation and Differentiation of Printed Skin Equivalents

1. Immediately after the overnight postprinting incubation, the printed skin equivalents are manually detached from the plate and grafted onto the backs of immunodeficient mice. To do this:

   (a) Full thickness circular wounds of 12 mm diameter are produced by means of a punch on the dorsum of each mouse.

   (b) Circular samples of the same diameter are obtained by the same punch from the printed skin substitutes, placed on the generated wounds (*see* **Note 11**), and covered by the skin, previously removed from these mice, devitalized by three cycles of freezing and thawing (*see* **Note 12**).

   (c) The devitalized skin is kept in place with the help of sutures for 1 month and a half.

   (d) The grafts are analyzed 8 weeks post grafting.

3.5 In Vitro Maturation and Differentiation of Printed Skin Equivalents

1. To form a multilayered epithelium with a well-developed stratum corneum, hKCs are allowed to differentiate at the air–liquid interface for 17 days at 37 °C in a CO_2 incubator in differentiating medium. The air–liquid interface is achieved by raising the transwell position within the culture plate, following the manufacturer's instructions.

2. The medium is changed every 3 days until day 17, in which the samples are removed from the transwells and analyzed.

3.6 Histology

1. Four biopsies generated either from in vivo or in vitro matured samples are collected with the help of a punch.

2. Samples are fixed using 3.7% buffered formaldehyde for 24 h and after that, they are washed three times with PBS and kept submerged in PBS for 24 h.

3. Samples are embedded in paraffin and cut into sections 5 μm wide with a microtome.

4. Samples are dewaxed, rehydrated, and stained with hematoxylin-eosin (H/E). (Fig. 2).

3.7 Immunostaining

1. Biopsies are embedded in Tissue-terk and frozen (-20 °C) for 24 h.

2. 5 μm wide cryosections are used for immunostaining for each of the primary antibodies (a total of 5).

3. Incubate samples in the air for 10–20 min at room temperature.

4. Wash the cryosections in PBS-(3%)BSA for 45 min at room temperature.

5. Add 30 μl of the corresponding primary antibody, diluted in PBS-(3%) BSA to the sections as specified below.
   Antibodies:
   (a) Antivimentin (ready to use).
   (b) Antikeratin 5 (K5) (1:1000).
   (c) Antikeratin 10 (K10) (1:100).
   (d) Antifilaggrin (1:200).
   (e) Anticollagen VII (1:5000).

6. Incubate samples for 1 h at 25 °C and then washed then three times in PBS for 5 min.

7. Add 30 μl of the corresponding secondary antibody, diluted 1:1000 in PBS-(3%) BSA to each cryosection. Incubate for 1 h at 25 °C and wash them three times in PBS 1× for 5 min (*see* **Note 13**).

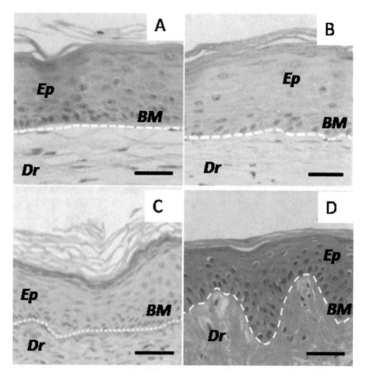

Fig. 2 (**a, b**). Histological analysis of 3D human skin equivalents generated in vitro after 17 days of differentiation at the air–liquid interface. (**a**) "Handmade" skin equivalent following our previous protocol. (**b**) Bioprinted skin equivalents. (**c**) Histological análisis (8 weeks post grafting) of bioprinted human skin grafted to immunodeficient mice. (**d**) Histological analisis of normal human skin. The white dotted line indicates the dermo-epidermal junction (basal membrane; BM). *Dr.* Dermis, *Ep* epidermis. Scale bar: 100 µm. (Modified with permission from [18])

8. Samples are coverslipped for image analysis (Fig. 3) using Mowiol mounting medium containing DAPI (1: 1000) for nuclei visualization (*see* **Notes 14** and **15**).

4 Notes

1. A more detailed description can be found in ref. 18.

2. Bioink 1 and Bioink 2 can be mixed in a single syringe to use only two syringes, two extruders and a bifurcated connector instead of three syringes, three extruders, and a trifurcated connector.

3. Optimize the length and diameter of the tubes connecting the syringes and the mixer. This will minimize the dead volume in the tubes and bioink waste.

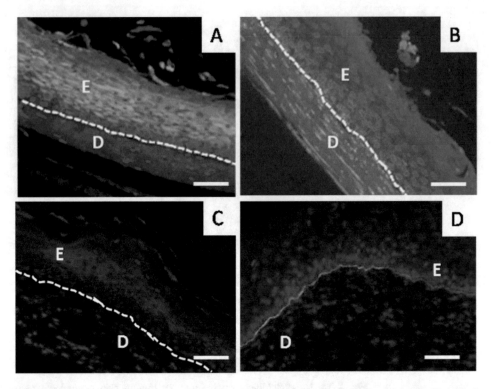

Fig. 3 Examples of immunostaining. (**a, b**) In vitro 3D human skin equivalents obtained after 17 days of differentiation at the air–liquid interface. Frozen sections stained (green) with the anti-K10 antibody, labeling differentiated suprabasal epidermal keratinocytes. (**b**) antihuman vimentin antibody (red), staining dermal human fibroblasts. (**c, d**) Immunohistochemical analysis (8 weeks post grafting) of bioprinted human skin grafted to immunodeficient mice. (**c**) Keratin K10 detection (red; suprabasal staining: notice that in **a** and **c** the basal layer is negative). (**d**) Collagen VII (green; dermo-epidermal junction) and vimentin (red; dermal compartment). *See* keratin K5, filaggrin, and smooth muscle actin detection in Fig. 5 in ref. 18. Scale bar: 100 μm. (*E* epidermis, *D* dermis, *dotted line* dermo-epidermal junction). (Modified with permission from [18])

4. You can prepare FM to culture 3T3 cells and to seed keratino-cytes, and after that, add to this culture medium 10 ng/ml of EGF in order to continue with keratinocyte culture. In that way, no culture medium is wasted.

5. Using a microscope, check keratinocytes morphology in order to avoid terminal differentiation.

6. To minimize the incubation time, wash gently with PBS prior Trypsin addition. Be careful when detaching the keratinocytes from culture surface. Do not harvest cell aggregates and sheets, only cells suspended individually.

7. Keratinocytes in suspension may overcome terminal differenti-ation. To avoid this, you can collect the keratinocytes and prepare bioink 4 while the dermal substitute is in the incubator. This will minimize the amount of time in which keratinocytes are suspended.

8. To minimize the effect of gravity-mediated precipitation, homogenize the cell-containing bioinks just before the printing process begins. This will ensure you evenly distributed cells in the printed skin.

9. Use the following flow rates to achieve a 12-ml/min flow rate of the mixture: 0.575 ml/min (bioink 1), 10.45 ml/min (bioink 2), and 0.975 ml/min (bioink 3).

10. Consider to program a routine to print consecutive wells in a 6-well plate.

11. The printed gel can be thinner and softer than expected. Manipulate it carefully to preserve its integrity.

12. When thawing, remove the skin from the bath immediately after the ice is melted in every cycle. Longer exposures to heat will result in skin contraction and deformation.

13. At this point, be careful with the light and maintain the samples in dark to avoid photobleaching of the fluorophore.

14. Leave the samples at room temperature and in the dark to allow the cover slips to attach to the glass slides and dry.

15. Samples can be stored at 4 °C in the dark (wrapped in aluminum) to prevent photobleaching for later analysis.

Acknowledgments

We kindly thank Prof. José Luis Jorcano (from UC3M) and Prof. Juan F. del Cañizo (Hospital General Universitario Gregorio Marañón) for their guidance and support. This work was supported by Programa de Actividades de I+D entre Grupos de Investigación de la Comunidad de Madrid, S2018/BAA-4480, Biopieltec-CM, Programa Estatal de I+D+i Orientada a los Retos de la Sociedad, RTI2018-101627-B-I00 and Cátedra Fundación Ramón Areces. The authors declare no conflict of interest.

References

1. Peng W, Unutmaz D, Ozbolat IT (2016) Bioprinting towards physiologically relevant tissue models for pharmaceutics. Trends Biotechnol 34(9):722–732

2. Holmes AM, Charlton A, Derby B, Ewart L, Scott A, Shu W (2017) Rising to the challenge: applying biofabrication approaches for better drug and chemical product development. Biofabrication 9(3):033001

3. Murphy SV, Atala A (2014) 3D bioprinting of tissues and organs. Nat Biotechnol 32:773–785

4. Arslan-Yildiz A, El Assal R, Chen P, Guven S, Inci F, Demirci U (2016) Towards artificial tissue models: past, present, and future of 3D bioprinting. Biofabrication 8(1):014103

5. Cui H, Nowicki MH, Fisher JP, Zhang LG, Zhang LG (2017) 3D bioprinting for organ regeneration. Adv Healthc Mater 6:1601118

6. Pereira RF, Sousa A, Barrias CC, Bayat A, Granja PL, Bartolo PJ (2017) Advances in bioprinted cell-laden hydrogels for skin tissue engineering. Biomanuf Rev 2:1

7. Ng WL, Wang S, Yeong WY, Naing MW (2016) Skin bioprinting: impeding reality or fantasy? Trends Biotechnol 34(9):689–699

8. Vijayavenkataraman S, Lu WF, Fuh JY (2016) 3D bioprinting of skin: a state-of-the-art review on modelling, materials, and processes. Biofabrication 8(3):032001

9. Velasco D, Quílez C, García M, Cañizo JF, Jorcano JL (2018) 3D human skin bioprinting: a view from the bio side. J 3D Print Med 2 (3):115–127

10. Augustine R (2018) Skin bioprinting: a novel approach for creating artificial skin from synthetic and natural building blocks. Prog Biomater 7:77–92. https://doi.org/10.1007/s40204-018-0087-0

11. Yan WC, Davoodi P, Vijayavenkataraman S, Tian Y, Ng WC, Fuh JYH, Robinson KS, Wang CH (2018) 3D bioprinting of skin tissue: from pre-processing to final product evaluation. Adv Drug Deliv Rev 132:270–295

12. Meana A, Iglesias J, Del Rio M, Larcher F, Madrigal B, Fresno MF, Martin C, San Roman F, Tevar F (1998) Large surface of cultured human epithelium obtained on a dermal matrix based on live fibroblast-containing fibrin gels. Burns 24:621–630

13. Llames SG, Del Rio M, Larcher F, García E, García M, Escamez MJ, Jorcano JL, Holguín P, Meana A (2004) Human plasma as a dermal scaffold for the generation of a completely autologous bioengineered skin. Transplantation 77:350–355

14. Llames S, García E, García V, del Río M, Larcher F, Jorcano JL, López E, Holguín P, Miralles F, Otero J, Meana A (2006) Clinical results of an autologous engineered skin. Cell Tissue Bank 7:47–53

15. Guerrero-Aspizua S, García M, Murillas R, Retamosa L, Illera N, Duarte B, Holguín A, Puig S, Hernández MI, Meana A, Jorcano JL, Larcher F, Carretero M, Del Río M (2010) Development of a bioengineered skin-humanized mouse model for psoriasis: dissecting epidermal-lymphocyte interacting pathways. Am J Pathol 177:3112–3124

16. García M, Llames S, García E, Meana A, Cuadrado N, Recasens M, Puig S, Nagore E, Illera N, Jorcano JL, Del Rio M, Larcher F (2010) In vivo assessment of acute UVB responses in normal and xeroderma pigmentosum (XP-C) skin humanized mouse models. Am J Pathol 177:865–872

17. Martínez-Santamaría L, Conti CJ, Llames S, García E, Retamosa L, Holguín A, Illera N, Duarte B, Camblor L, Llaneza JM, Jorcano JL, Larcher F, Meana Á, Escámez MJ, Del Río M (2013) The regenerative potential of fibroblasts in a new diabetes-induced delayed humanized wound healing model. Exp Dermatol 22:195–201

18. Cubo N, Garcia M, Del Cañizo JF, Velasco D, Jorcano JL (2017) 3D bioprinting of functional human skin: production and in vivo analysis. Biofabrication 9(1):015006

19. Walker RH (ed) (1993) Technical manual, 11th edn. American Association of Blood Banks, Bethesda, MD, pp 728–730

Chapter 15

3D Bioprinting and Differentiation of Primary Skeletal Muscle Progenitor Cells

Catherine Ngan, Anita Quigley, Cathal O'Connell, Magdalena Kita, Justin Bourke, Gordon G. Wallace, Peter Choong, and Robert M. I. Kapsa

Abstract

Volumetric loss of skeletal muscle can occur through sports injuries, surgical ablation, trauma, motor or industrial accident, and war-related injury. Likewise, massive and ultimately catastrophic muscle cell loss occurs over time with progressive degenerative muscle diseases, such as the muscular dystrophies. Repair of volumetric loss of skeletal muscle requires replacement of large volumes of tissue to restore function. Repair of larger lesions cannot be achieved by injection of stem cells or muscle progenitor cells into the lesion in absence of a supportive scaffold that (1) provides trophic support for the cells and the recipient tissue environment, (2) appropriate differentiational cues, and (3) structural geometry for defining critical organ/tissue components/niches necessary or a functional outcome. 3D bioprinting technologies offer the possibility of printing orientated 3D structures that support skeletal muscle regeneration with provision for appropriately compartmentalized components ranging across regenerative to functional niches. This chapter includes protocols that provide for the generation of robust skeletal muscle cell precursors and methods for their inclusion into methacrylated gelatin (GelMa) constructs using 3D bioprinting.

Key words 3D bioprinting, Skeletal muscle, Myoblasts, Tissue engineering

1 Introduction

Skeletal muscle comprises approximately 40% of the total body mass and is critical for day to day function. Damage and/or loss of skeletal muscle occurs through diseases such as the muscular dystrophies and trauma such as sports injury or motor accident. While minor localized injuries in skeletal muscle can repair on their own through the proliferation and fusion of local progenitor cells to form new fibers or to repair damaged fibers [1, 2], large muscle defects cannot be repaired endogenously.

Significant volumetric muscle loss is generally addressed by surgical intervention to replace lost muscle with autologous muscle harvested from near the site of injury with its own nerve and vascular supply [3]. However, there is significant donor site

Jeremy M. Crook (ed.), *3D Bioprinting: Principles and Protocols*, Methods in Molecular Biology, vol. 2140,
https://doi.org/10.1007/978-1-0716-0520-2_15, © Springer Science+Business Media, LLC, part of Springer Nature 2020

morbidity associated with these procedures and there is variable success in the outcome of this type of surgery [4]. As such, there is a need for tissue engineering approaches that can meet the challenges associated with the replacement and regeneration of significant areas of muscle, not only for trauma-derived volumetric muscle loss injuries but also for degenerative muscle disease involving skeletal muscle atrophy and/or dystrophy.

One major issue with tissue engineering large areas of ablated tissues is restoration of functional 3D tissue structure. This is a major challenge for muscle replacement as skeletal muscle is a highly organized tissue, with defined anisotropic properties. Muscle tissues require structured, three-dimensional (3D) orientation and alignment of mature muscle fibers, as well as extensive innervation and vascularization, and this has presented a challenge in the ex vivo engineering of skeletal muscle. Developments in 3D bioprinting present an exciting step forward in overcoming some of these challenges. 3D bioprinting allows the engineering of structured, orientated, cell-laden scaffolds that can be constructed to contain multifunctional properties. Constructs can be engineered to contain different cell types, strategically placed growth factors, as well provision for vascularization and innervation, so that regeneration can occur in a controlled fashion [5–7].

Our laboratories have been developing bioink and bioprinting methods for production of viable 3D printed skeletal muscle constructs to address some of the issues associated with the ex vivo engineering of skeletal muscle [8, 9]. This communication describes our protocol for the characterization of myoblast populations suitable for generating 3D muscle constructs and 3D bioprinting and for the differentiation of skeletal muscle progenitor cells using methacrylated gelatin (GelMA).

2   Materials

2.1 Characterization of Primary Myoblasts

1. Primary murine myoblasts, 40×10^6 cells are required for the current protocol (20 million cells required per mL of bioink solution (*see* Subheading 3.2)).

2. Tissue culture flasks, 175 cm^2 (Corning).

3. Myoblast proliferation media: Hams F-10 (Gibco), 20% fetal bovine serum (FBS, Gibco), 2.5 ng/mL recombinant human basic fibroblast growth factor (bFGF, Peprotec), 2 mM L-glutamine (Gibco), 100 U/mL penicillin, and 100 μg/mL of streptomycin (Gibco).

4. Dissociation buffer (8.5 mM NaCl, 0.5 mM KCl, 2.3 mM NaHCO$_3$, 0.8 mM NaH$_2$PO$_4$·2H$_2$O, 0.56 mM Glucose, 0.096 mM EDTA, 10 mg/L phenol red, adjust pH to 7.4 prior to addition of trypsin (Life Technologies) to 0.25%).

5. Flow cytometer (e.g., Cytoflex, Beckman Coulter) and analysis software (e.g., FlowJo v10.0).

6. Flow cytometry buffer (1× PBS, 0.5% BSA, 2 mM EDTA, sterile filtered).

7. AF647 conjugated mouse antidesmin (Santa Cruz Biotechnology).

8. AF488 conjugated anti-Pax7 (Novus Biologicals).

9. APC conjugated anti-α7 integrin (Miltenyi Biotech).

10. Isotype controls.

11. Benchtop centrifuge.

12. Flow cytometry tubes.

2.2 Sterilization and Preparation of GelMA

1. Dry, nonsterile methacrylated gelatin (GelMA; provided by the Australian Fabrication Facility (ANFF), University of Wollongong NSW). Sterile GelMA is available commercially from Cellink.

2. Sterile mini bioreactor tubes with cent cap (50 mL) (Corning).

3. Sterile 50 mL tubes (Corning).

4. Sterile syringe barrels, Luer lock (50 mL volume, BD Biosciences).

5. Syringe filters (0.22 μm pore size, Millipore).

6. Cell strainers (40 μm, BD Biosciences).

7. Dry, sterile (autoclaved) 1000 mL Schott bottles (×2).

8. Glass beaker.

9. Analytical balance.

10. Freeze dryer.

11. Class II biological safety cabinet.

12. −80 °C freezer.

2.3 Encapsulation of Myoblasts in GelMA and Transfer of Ink to Print Cartridge

1. Myoblasts in growth phase (enough for a final concentration of approximately 20×10^6 cells/mL of material).

2. Myoblast growth media (*see* above).

3. Dry sterile GelMA (with known mass).

4. Sterile PBS.

5. Penicillin (100 U/mL) and streptomycin (100 μg/mL) (Gibco).

6. Sterile 2% w/v stock of lithium phenyl-2,4,6-trimethylbenzoylphosphinate (LAP, Tokyo Chemical Industry Co., Ltd.) photoinitiator, in PBS.

7. Dissociation buffer (Subheading 2.1, **item 4**).

8. Sterile 50 mL conical centrifuge tubes.

9. Sterile 1 mL and 200 μL barrier pipette tips.

10. 1 mL sterile, individually wrapped transfer pipettes (Copan).

11. Sterile ink cartridges with piston, tip cap, and conical nozzles (27 G) (Nordson).

12. Aluminum foil.

13. Trypan blue (Lonza).

14. Cell counter, or hemocytometer.

15. Benchtop centrifuge.

16. Benchtop vortex.

17. Class II biosafety cabinet.

18. 4 °C fridge.

19. Tissue culture incubator (humidified, 37 °C, 5% CO_2).

20. Standard tissue culture microscope.

2.4 The Printing Process

1. Sterile 6-well tissue culture plates.

2. Petri dish.

3. Commercial pneumatic extrusion printer system with isolation chamber (e.g., Inkredible+ 3D bioprinter by Cellink).

4. Myoblast growth media.

5. 365 nm UV light source (e.g., Omnicure LX400, Lumen Dynamix LDGI).

6. Tissue culture incubator (humidified, 37 °C, 5% CO_2).

2.5 Differentiation and Analysis of 3D Printed Muscle Constructs

1. Myoblast differentiation media: DMEM, 2% horse serum, 2 mM L-glutamine, 100 U/mL penicillin, and 100 μg/mL streptomycin.

2. Tissue culture incubator (humidified, 37 °C, 5% CO_2).

3. Sterile, 1 mL barrier pipette tips.

4. Sterile transfer pipettes.

5. Calcium- and magnesium-free 1× PBS (Lonza).

6. 10% neutral buffered formalin (Sigma Aldrich).

7. 0.1% Triton X-100 (Sigma Aldrich) in PBS.

8. 1% bovine serum albumin (BSA, Sigma Aldrich) in sterile calcium and magnesium free 1× PBS (Lonza).

9. Alexa Fluor 488 Phalloidin (Life Technologies) at 6.6 μM in methanol.

10. 4′,6-Diamidino-2-phenylindole dihydrochloride (DAPI, Life Technologies).

11. Standard tissue culture microscope.

12. Inverted fluorescent microscope.

13. Confocal microscope (e.g., Nikon A1R).

3 Methods

3.1 Myoblast Preparation

For the purpose of this protocol, primary myoblasts were purified from the hind limbs of a 5-week-old male C57Bl/6 mouse essentially as described elsewhere [10] (*see* **Note 1**). Details of the procedure used in this protocol are beyond the scope of the current chapter, however suitable methods can be found in many publications [10–12]. Alternatively, the C2C12 cell line can be used as a model for primary muscle cells. When working with primary cultures, the quality of the preparation is of paramount importance to ensure efficient maturation and differentiation to myotubes. Our laboratory routinely assesses the purity of myogenic cultures by flow cytometry for myogenic markers (α7 integrin, Pax7, and desmin) before experiments are carried out (Fig. 1). This is particularly important, particularly when testing the efficacy of biomaterials to support cell differentiation. We have included our protocol for standard flow cytometry in this chapter as a guide for assessing the primary myogenic cultures suitable for printing and differentiation.

3.2 Flow Cytometry of Primary Myoblasts

While some antigens can withstand the fixation process, staining of some cell surface antigens requires the use of "nonfixed" cells to preserve antigen morphology. In addition, for cell surface antigens, permeabilization of the membrane with detergents such as Triton x-100 is not generally required. Optimization of staining procedures needs to be performed on a per user basis for each antigen. For staining of α7 integrin with the antibody used in this protocol (Miltenyi Biotech), we recommend the use of freshly isolated, nonfixed, nonpermeabilized cells to achieve optimal results. This can be achieved using the above protocol with the omission of **steps 2** and **3** (fixation) and the omission of triton X-100 from the blocking solution in **step 4**. All incubations with fresh cells should be carried out on ice (4 °C), with chilled flow cytometry buffer, to ensure maximum viability of the analyzed cells.

3.2.1 Flow Cytometry of Myogenic Precursors

1. Myoblasts are collected from tissue culture plates with dissociation buffer and resuspended in cell media. Ensure myoblasts are a single cell suspension and there are no clumped cells. A concentration of 1×10^6 cells/mL is optimal for staining.

2. Fix cells in 10% formalin for 15 min at room temperature (RT; omit for α integrin staining).

3. Centrifuge cells to a pellet (800 × g, 5 min) and remove formalin solution (discard into chemical waste container according to institutional protocols).

4. Wash cells twice in flow cytometry buffer, and repellet by centrifugation.

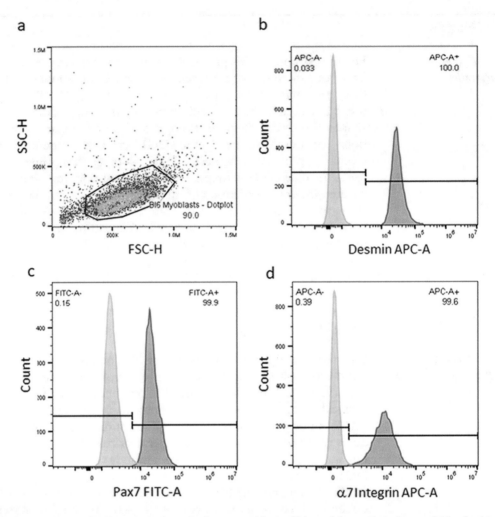

Fig. 1 Flow cytometric characterization of primary myoblasts. Side and forward scatter analysis of primary myoblast cultures reveals a single population of cells (**a**). Myoblasts progenitors are predominantly desmin (**b**), Pax7 (**c**), and α7-integrin (**d**) positive. Control cell populations are shaded blue, while stained populations are shaded red

5. Resuspend cells in blocking buffer containing 0.1% triton X-100, for permeabilization of the plasma and nuclear membranes, for 1 h at 4 °C (omit triton X-100 from buffer for α7 integrin staining).

6. Pellet the cells by centrifugation ($800 \times g$, 5 min) and washed cell pellet once in flow buffer by centrifugation. Resuspend the cell pellet at 1×10^6 cells/mL.

7. Aliquot 100 μL of the cell suspension into separate flow cytometry tubes for staining, one tube per antibody/isotype control. An unstained control should also be included to determine the extent of nonspecific isotype binding.

8. Incubate cells with fluorescently labeled primary antibodies or isotype controls. Use the antibodies at recommended concentrations or at concentrations predetermined by the user. An unstained control should also be included to determine the extent of nonspecific isotype binding. Incubate the cells for 1 h at 4 °C.

9. After incubation, add 500 µL of chilled flow cytometry buffer to each tube and centrifuge the cells to a pellet ($800 \times g$, 5 min).

10. Remove the supernatant and wash each cell pellet with a further 500 µL of chilled flow cytometry buffer.

11. Resuspend each pellet in 100 µL flow cytometry buffer for analysis. Cells should be at a final concentration of approximately ($0.5-1 \times 10^6$ cells/mL).

3.3 Sterilization of GelMA

Gelatin methacrylate (GelMA) can be synthesized as previously described [13] (*See* **Note 9**). The sterilization process takes approximately 1 week to complete. Alternatively, sterile GelMA is available commercially from Cellink (Sweden) (*see* **Note 2**).

Part 1

1. Measure and record the mass of one dry, sterile 1000 mL Schott bottle.

2. Place the bottle on a scale and add dry GelMA to the bottle up to a maximum mass of 15 g (*see* **Note 3**).

3. Record the total mass of the bottle and material.

4. Add MilliQ water to the GelMA to a final concentration of 1.5% w/v GelMA.

5. Place on rotating platform in an incubator at 37 °C for 2–3 h (or more) until total dissolution of GelMA. Cover in foil to protect from ambient light.

6. In the meantime, autoclave a clean 1000 mL Schott bottle.

7. In a class II biological safety cabinet (with the light off), pour the warm, dissolved GelMA solution through a cell strainer into a glass beaker for drawing up into a sterile 50 mL syringe.

8. After drawing up the GelMA solution, attach a 0.2 µm syringe filter to the end (maintain sterility). Note: you will need a number of sterile 0.2 µm filters for this process, depending on the viscosity of the GelMA solution.

9. Filter sterilize the material into the sterile autoclaved Schott bottle.

10. After sterilizing all of the material, keep it warm and protected from light at 37 °C for ease of handling.

11. To prepare for the final aliquots of material, measure and record the mass of each empty sterile vented-capped 50 mL centrifuge.

12. Add 40 mL (or volume required for a specific mass if required) of GelMA solution to each centrifuge tube.

13. Place in −80 °C freezer overnight.

Part 2

1. Transfer all aliquots to the freeze-dryer without allowing the solutions to thaw.

2. Freeze-dry until the GelMA is fully dehydrated (typically 4–7 days).

3. To check if the process is complete, take a repeated mass measurement of an aliquot several hours apart. If the mass does not change, the drying process is complete.

4. Measure and record the mass of the dried GelMA and centrifuge tubes. Subtract the mass of the original empty tube to calculate the mass of dry GelMA.

5. In a biosafety cabinet, exchange the vented caps with standard screw-top caps.

6. Store material at −80 °C until required.

3.4 Encapsulation of Myoblasts in GelMA

The volumes in this protocol depend on the required volume of bioink and the desired final concentration of GelMA. The following protocol will make 2 mL of bioink with 8% wt/vol GelMA and 0.1% LAP, at a cell density of 20×10^6 cells/mL. These values can be adjusted according to user requirements. All work should be performed in a class II biosafety cabinet.

Part 1

1. Prepare a stock solution of 20% wt/vol GelMA. To do this, use the mass of the dry hydrogel, calculated from the sterilization process, to calculate the volume of sterile PBS required. It is optional to include antibiotics (penicillin/streptomycin) at this stage.

2. Dissolve at 37 °C on a shaker. This may take several hours.

3. Once fully dissolved, store material at 4 °C.

Part 2

1. To prepare the myoblasts, transfer the culture media from the tissue culture flask to a 50 mL centrifuge tube.

2. Gently rinse the cells with sterile PBS and transfer this to the 50 mL centrifuge tube.

3. Add enough dissociation buffer to cover the culture surface and incubate at 37 °C for 3–5 min.

4. Observe the cells under the microscope. When detached, transfer the solution to the same 50 mL centrifuge tube.

5. Centrifuge cells to a pellet at 800 × g for 5 min.

6. Remove the supernatant, and gently resuspend the cell pellet in 1–5 mL of growth medium, depending on the cell density.

7. Count the cells with a cell counter/hemocytometer, using Trypan blue to stain the cells.

8. Aliquot the appropriate volume of cell suspension for the number of cells required in the final bioink composition into a new centrifuge tube. In this example, 40 million cells would be required per 2 mL of GelMA solution.

9. Centrifuge this aliquot at 800 × g for 5 min. Discard the supernatant.

10. Resuspend the cell pellet in 1100 µL of myoblast growth media.

11. Warm up the stock 20% wt/vol GelMA for 5–10 min at 37 °C. Add 800 µL of warmed 20% wt/vol stock GelMA solution to the cell suspension, and 100 µL of 2% wt/vol LAP stock for a final concentration of 0.1% wt/vol. Be wary of unnecessary light exposure to the sample now that the photoinitiator has been added. Cover the tube with foil.

12. Very gently vortex the cell suspension. Rewarm to 37 °C if required.

13. To prepare the ink cartridge, make sure the tip cap is firmly on the cartridge. With a 1 mL pipette, transfer the GelMA myoblast ink to the cartridge. Insert the piston at the other end, then carefully invert the cartridge.

14. Remove the tip cap, and slowly advance the piston to prime the cartridge and minimize the deadspace at the tip. Reattach the tip cap.

15. Transfer the cartridge to a 4 °C fridge for 20 min. Also place the 6-well plate to be used for printing and the conical nozzles in the fridge to cool.

3.5 The Printing Process

The shape of the printed construct requires g-code programming to instruct the bioprinter as to the required geometry and sequence of events (*see* **Note 10**). The specifics of g-code are beyond the scope of the current protocol which assumes that a code is available and optimized for generating the required bioprinted construct (*see* Subheading 4 for more details). The current protocol describes the process followed in our laboratories for bioprinting with a CellInk Inkredible Plus bioprinter (Fig. 2) (*see* **Note 4**).

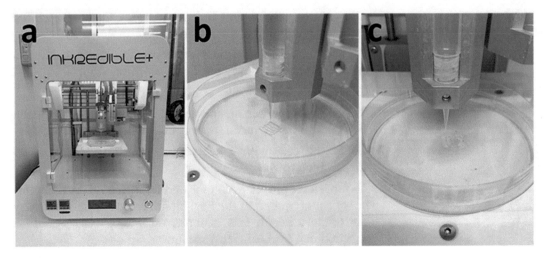

Fig. 2 Setup of the Celllnk Inkredible plus printer (**a**), printed grid construct (**b**), and exposure of printed construct to cross-linking UV light (365 nm) for 400 s (4 mW/cm^2) (**c**)

1. Turn on the bioprinter and air compressor (*see* **Note 5**).

2. Take the cartridge and nozzles out of the fridge. In the biosafety hood, remove the tip cap and attach the nozzle. Prime the nozzle by gently applying pressure to the piston, so that the GelMA bioink fills the cone, and there is minimal deadspace.

3. Attach the air tube adapter to the cartridge and lock the cartridge into the printhead holder (PH1).

4. Test the air pressure to determine the optimal extrusion speed. Using the pressure regulator wheels on the printer, turn the pressure down to zero. Using the LCD screen, under the "Prepare Bioprint" menu, select PH1 "on." Gradually increase the pressure until there is a steady extrusion of bioink, which can be collected on a spare petri dish and discarded. Turn PH1 off (*see* **Note 6**).

5. Place the cold 6-well plate onto the stage.

6. Calibrate the X and Y axis by selecting the "Home Axes" option in the printer LCD menu.

7. Calibrate the Z axis to the bottom of the plate by selecting "Move Z" and scrolling to the desired height. Then select "Calibrate Z."

8. In the LCD screen, find the file to print under the "Bioprint" menu.

9. Adjust the pressures as required during the printing process.

10. After printing, immediately expose the construct to the UV light source, adjusted to 4 mW/cm^2 intensity. Cure for 400 s, or whatever time is required for cross-linking of the material as previously determined by rheology testing (*see* **Note 7**).

11. After cross-linking, rinse the constructs in myoblast growth media.

12. Cover the constructs with myoblast growth media, and place in a tissue culture incubator.

3.6 Differentiation and Analysis of 3D Printed Muscle Constructs

1. After 24 h, change the myoblast growth media to differentiation media.

2. Continue with daily half-media changes. Early myotube formation should be visible with bright field microscopy within a few days.

3. By day 3 of differentiation, the constructs can be gently eased off the tissue culture surface to be suspended in the media. This is to encourage even diffusion of nutrients.

4. Extensive myotube formation should be observed between days 7 and 14 of differentiation.

Constructs can be fixed and stained for imaging with fluorescent microscopy, and further characterized with 3D-rendered z-stacks with confocal imaging. Confocal staining allows the visualization of cell distribution within the printed constructs (Fig. 3).

3.7 F-Actin Staining

1. Wash constructs in PBS (*see* **Note 8**).

2. Fix with 4% PFA for 30 min.

3. Wash with PBS.

4. Permeabilize cells by covering the constructs with 0.1% Triton X-100 in PBS for 1 h.

5. Wash with PBS. Allow the constructs to soak in PBS for 30 min to minimize background staining of the material.

6. Stain with a 1 in 100 dilution of Alexa Fluor 488 Phalloidin in 1% BSA. Leave covered in foil for 1–2 h at RT.

7. Wash with PBS, again leaving for at least 30 min to wash out the stain.

8. Stain cell nuclei for 15 min at RT with 0.1 µg/mL DAPI diluted in PBS.

9. Image with standard fluorescent microscope or confocal microscope for 3D-rendered images.

4 Notes

1. When working with primary cells, it is generally a good idea to check their capacity to fuse and generate myotubes prior to printing. The majority of our experiments are carried out with primary mouse myoblast cultures at low passage number; we find these cultures generally have a fusion index (% of nuclei within myotubes) of approximately 80%.

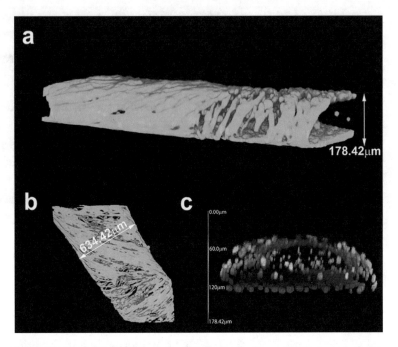

Fig. 3 Confocal 3D rendering of printed myogenic progenitors subjected to differentiation conditions for 7 days. The 3D printed lattice consisted of fibers approximately 178 μm in height (**a**) and 634 μm in width (**b**). Extensive myotube formation can be observed on the surface of the printed scaffold (phalloidin-AF488 stained), while myogenic nuclei (DAPI labeled) can be observed both on the surface and within the cross-lined GelMA, as illustrated by depth coding (**c**)

2. When filter sterilizing GelMA, a lower % GelMA solution may increase ease of filtering, however there will be less material per falcon tube at the end of the freeze drying process.

3. When working with GelMA and/or LAP, work in a biological safety cabinet at all times to maintain sterility, with solutions protected from light. In our laboratories, we work with the biological safety cabinet light turned off to reduce incidental light exposure.

4. Sample g-code files for bioprinting are available on most bioprinter manufacturer websites (e.g., https://cellink.com/support/) and commonly come preloaded on most commercially available bioprinters. There are a number of freeware programs available for visualizing and editing g-code files, our laboratories use various g-code editors including Repetier (www.repetier.com).

5. Cool cell loaded GelMA and collection plate to four degrees before proceeding with bioprinting. This helps maintain shape fidelity prior to crosslinking.

6. Note that print speeds of about 200 mm/min through a 27 G nozzle can achieve a fiber width of 200–250 μm. It is important to monitor and adjust printing pressure while printing to achieve optimal results.

7. Prior to crosslinking, cover reagents containing LAP with foil to protect from light.

8. For confocal microscopy, we remove the fixed and stained constructs from their plates and place in a glass bottomed dish, suitable for confocal microscopy. The constructs are hydrated with a minimal volume of PBS. Minimize the volume of PBS in the dish to ensure that the printed construct remains fixed in place while imaging.

9. A number of different bioinks are commercially available through various companies including CellInk (https://cellink.com) and Bioink Solutions (https://www.bioinksolutions.com) among others.

10. We recommend optimization of the printing process with bioink of interest prior to bioprinting cells. Other chapters in this book provide protocols for optimizing bioprinting parameters.

Acknowledgements

The authors would like to thank the Australian National Fabrication Facility (ANFF), the Australian Research Council (CE140100012), and MTP Connect for supporting this work.

References

1. Nag AC, Foster JD (1981) Myogenesis in adult mammalian skeletal muscle in vitro. J Anat 132 (Pt 1):1–18

2. Snow MH (1977) Myogenic cell formation in regenerating rat skeletal muscle injured by mincing. I. A fine structural study. Anat Rec 188(2):181–199. https://doi.org/10.1002/ar.1091880205

3. Garg K, Ward CL, Hurtgen BJ, Wilken JM, Stinner DJ, Wenke JC et al (2015) Volumetric muscle loss: persistent functional deficits beyond frank loss of tissue. J Orthop Res 33 (1):40–46. https://doi.org/10.1002/jor.22730

4. Grogan BF, Hsu JR, Skeletal Trauma Research C (2011) Volumetric muscle loss. J Am Acad Orthop Surg 19(Suppl 1):S35–S37

5. Holmes B, Bulusu K, Plesniak M, Zhang LG (2016) A synergistic approach to the design, fabrication and evaluation of 3D printed micro and nano featured scaffolds for vascularized bone tissue repair. Nanotechnology 27 (6):064001. https://doi.org/10.1088/0957-4484/27/6/064001

6. Lee YB, Polio S, Lee W, Dai G, Menon L, Carroll RS, Yoo SS (2010) Bio-printing of collagen and VEGF-releasing fibrin gel scaffolds for neural stem cell culture. Exp Neurol 223 (2):645–652. https://doi.org/10.1016/j.expneurol.2010.02.014

7. Yurie H, Ikeguchi R, Aoyama T, Kaizawa Y, Tajino J, Ito A, Ohta S, Oda H, Takeuchi H, Akieda S, Tsuji M, Nakayama K, Matsuda S (2017) The efficacy of a scaffold-free Bio 3D conduit developed from human fibroblasts on peripheral nerve regeneration in a rat sciatic nerve model. PLoS One 12(2):e0171448. https://doi.org/10.1371/journal.pone.0171448

8. Chung JHY, Naficy S, Yue ZL, Kapsa R, Quigley A, Moulton SE et al (2013) Bioink properties and printability for extrusion printing living cells. Biomater Sci 1(7):763–773. https://doi.org/10.1039/c3bm00012e

9. Ferris CJ, Stevens LR, Gilmore KJ, Mume E, Greguric I, Kirchmajer DM et al (2015) Peptide modification of purified gellan gum. J Mater Chem B 3(6):1106–1115. https://doi.org/10.1039/c4tb01727g

10. Todaro M, Quigley A, Kita M, Chin J, Lowes K, Kornberg AJ et al (2007) Effective detection of corrected dystrophin loci in mdx mouse myogenic precursors. Hum Mutat 28 (8):816–823. https://doi.org/10.1002/humu.20494

11. Lavasani M, Lu A, Thompson SD, Robbins PD, Huard J, Niedernhofer LJ (2013) Isolation of muscle-derived stem/progenitor cells based on adhesion characteristics to collagen-coated surfaces. Methods Mol Biol 976:53–65. https://doi.org/10.1007/978-1-62703-317-6_5

12. Motohashi N, Asakura Y, Asakura A (2014) Isolation, culture, and transplantation of muscle satellite cells. J Vis Exp (86). https://doi.org/10.3791/50846

13. O'Connell CD, Di Bella C, Thompson F, Augustine C, Beirne S, Cornock R et al (2016) Development of the Biopen: a hand-held device for surgical printing of adipose stem cells at a chondral wound site. Biofabrication 8(1):015019. https://doi.org/10.1088/1758-5090/8/1/015019

Chapter 16

Bioprinting Strategies for Secretory Epithelial Organoids

Ganokon Urkasemsin, Sasitorn Rungarunlert, and João N. Ferreira

Abstract

Novel three-dimensional (3D) biofabrication platforms can allow magnetic 3D bioprinting (M3DB) by using magnetic nanoparticles to tag cells and then spatially arrange them in 3D around magnet dots. Here, we report an M3DB methodology to generate salivary gland-like epithelial organoids from stem cells. These organoids possess a neuronal network that responds to saliva neurostimulants.

Key words Bioprinting, Magnetic nanoparticles, Organoids, Salivary gland, Xerostomia

1 Introduction

Human secretory glands, such as mammary, prostate, thyroid, pancreas, and salivary glands, are organs frequently targeted by cancer epidemics worldwide [1]. Functional damage in these glands usually occurs after conventional cancer therapies are delivered, thereby impairing the health of patients. Immunosuppression-free gland transplants are currently not available for gland replacement. Organoid-based bioprinting has been emerging as a promising biofabrication strategy to promptly develop an in vitro and "off-the-shelf" autologous gland transplant from patient's own cryopreserved cells [2, 3]. To avoid immunological incompatibilities and potential cytotoxicity from scaffold biomaterials in the long term, our group generated a scaffold-free culture platform to promptly produce 3D functional salivary gland-like organoids, which was validated in our previous works [4, 5]. This M3DB biofabrication process induces cellular spatial arrangement in 3D and scalable spheroid production within 24 h. In addition, this process facilitates spheroid handling and transfer, as well as high-throughput analysis and imaging. A magnetic nanoparticle solution consisting of gold, iron oxide, and poly-L-lysine is used in this biofabrication. These nanoparticles support cell proliferation and metabolism, without increasing deleterious processes such as inflammation and

Jeremy M. Crook (ed.), *3D Bioprinting: Principles and Protocols*, Methods in Molecular Biology, vol. 2140,
https://doi.org/10.1007/978-1-0716-0520-2_16, © Springer Science+Business Media, LLC, part of Springer Nature 2020

oxidative stress [6]. In addition, negligible immune response after transplantation is reported in animal models [7].

Thus, in this chapter, we describe the methodology to generate a secretory epithelial organoid similar to the native salivary gland (SG) using stem cells and an M3DB platform in vitro. The overall objective was to generate a scalable SG organoid with innervation and biofunctional properties upon neurostimulation to overcome current limitations in stem cell therapies for hypofunctional SG [8].

2 Materials

2.1 Cell Lines

1. Human dental pulp stem cells (hDPSC) (*see* **Note 1**).

2.2 Chemicals and Culture-Grade Reagents

1. High-glucose Dulbecco's modified Eagle medium (DMEM, Invitrogen, cat. no. 11965-092).

2. Fetal bovine serum (FBS, Gibco, cat. no. 26140079) (*see* **Note 2**).

3. GlutaMAX™ Supplement (Invitrogen, cat. no. 35050-061).

4. Retinoic acid (Sigma, cat. no. R2625).

5. Defined Keratinocyte Media—SFM 1X (Gibco, cat. no. 10744019).

6. Human fibroblast growth factor 10 (FGF-10) protein (R&D systems, cat. no. 345-FG-025).

7. Penicillin/streptomycin (Invitrogen, cat. no. 15140-122).

8. TrypLE Select Enzyme (1×), no phenol red (Gibco, cat. no. 12563011).

9. Trypan blue solution (0.4% (wt/vol), Gibco, cat. no. 15250061).

10. Phosphate-buffered saline without Ca^{2+}/Mg^{2+} (PBS, Invitrogen, cat. no. 10010-023).

11. NanoShuttle™-PL solution (Nano3D Biosciences, cat. no. 005-NS).

12. CellTiter-Glo® 3D cell viability solution (Promega Corporation, cat. no. G9681).

13. EnzChek™ *Ultra* Amylase Assay Kit (Invitrogen, cat. no. E33651).

2.3 Disposables

1. 75 cm^2 cell culture-treated Nunc™ EasYFlask™ flasks (Thermo Scientific™, ThermoFisher Scientific, cat. no. 156499).

2. Ultra-low attachment 96-well or 384-well clear flat-bottom plate (Corning® Costar®, Sigma, cat. nos. CLS3474 and CLS3827XX1).

3. Conical tubes (15 mL; Falcon, cat. no. 352097).

4. Serological pipettes (5, 10, and 25 mL; Thermo Scientific Nunc, cat. nos. 170355, 170356, and 170357).

2.4 Equipment

1. Inverted *xyz*-motorized phase-contrast microscope (Leica DMi8 microscope with automated *xy* platform and *z*-staking, Leica Microsystems, Germany).

2. Perfection V550 Epson flatbed scanner (Seiko Epson Corporation, Japan).

3. Biological safety cabinet (Class A2, ESCO).

4. Humidified incubator (37 °C, 5% CO_2, ESCO).

5. Centrifuge (Eppendorf, 5804R).

6. Hemocytometer, Neubauer improved (Carl Roth, cat. no. T728.1).

7. Magnetic drives of 96 or 384 neodymium magnet pins (Nano3D Biosciences, cat. nos. 011-96WK and 012-384WK).

8. Micropipette (1–10, 10–100, and 100–1000 μL, Eppendorf, cat. nos. 3121000023, 3121000074, and 3121000120).

9. Glomax® luminometer (Promega Corporation).

10. Fluorescence-based microplate reader (Tecan Group Ltd.).

2.5 Reagent Preparation

1. hDPSC growth medium for monolayer cultures: DMEM containing 10% (vol/vol) FBS, 2 mM GlutaMAX™, and 1% penicillin/streptomycin (50 μg/mL streptomycin and 50 U/mL penicillin). Filter sterilize and store at 4 °C for up to 2 weeks.

2. hDPSC growth medium for 3D spheroid formation cultures for epithelial progenitor enrichment: DMEM containing 10% (vol/vol) FBS, 2 mM GlutaMAX, and 1% penicillin/streptomycin (50 μg/mL streptomycin and 50 U/mL penicillin) supplemented with 5 μM retinoic acid. Filter sterilize and store at 4 °C for up to 2 weeks (*see* **Note 3**).

3. Epithelial differentiation medium: Defined Keratinocyte Media—SFM 1X supplemented with FGF-10 (4–400 ng/mL). Store at 4 °C for up to 2 weeks (*see* **Note 4**).

3 Methods (Fig. 1)

Set up all cell culture experiments within a sterilized biological safety cabinet. Warm up all media to 37 °C in a water bath. Change medium daily during 3D spheroid formation stage (3 days) and every 2 days during 2D monolayer expansion and 3D epithelial differentiation stage (8 days). All cell culture flasks and plates were kept at 37 °C in a humidified incubator containing 5% CO_2.

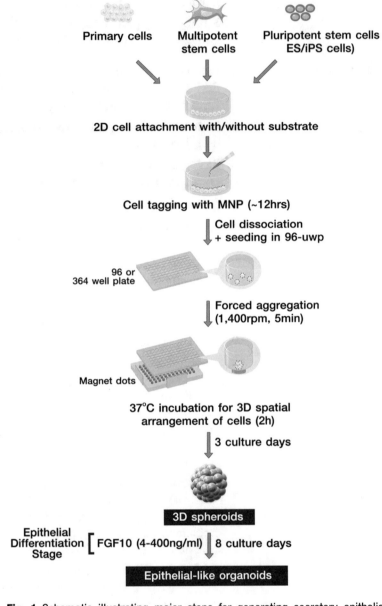

Fig. 1 Schematic illustrating major steps for generating secretory epithelial organoids

3.1 hDPSC
Expansion
in Monolayer Cultures

1. Culture hDPSC in flasks for at least two passages after thawing in hDPSC growth medium for monolayer cultures and after reaching 75–80% confluency (*see* **Note 5**).

2. Use Trypan blue exclusion method to count the number of viable cells between passages and before final seeding in 3D culture platform.

3.2 Transfer to M3DB Culture Platform for 3D Spheroid Formation Stage

1. Incubate confluent hDPSC (75–80%) overnight with NanoShuttle™-PL solution at a concentration of 1 μL/2.5×10^4 cells to tag cells with magnetic properties at plasma membrane.

2. Gently wash magnetized cells with PBS, and enzymatically detach them with TrypLE solution.

3. Resuspend cells in growth media for 3D spheroid formation.

4. Seed resuspended cells into ultralow attachment plates at 3×10^6 cells/well (for 96-well plates) or 1×10^6 cells/well (for 384-well plates).

5. Centrifuge plate by forced aggregation at $184 \times g$ for 5 min to facilitate cell clustering at the bottom of the well.

6. Place magnet dots (magnetic drive) below the plates to assemble the cells into spheroids at the bottom of the well.

7. Place the plates together with the magnet dots in the CO_2 incubator for at least 2 h.

3.3 3D Spheroid Formation Stage in M3DB

1. Remove the magnet dots, and change media every day for 3 days during 3D spheroid formation stage.

2. Scan plate with the Epson flatbed scanner to quantify the diameter size of the spheroids every day. This diameter should decrease with increased cell viability, and cell–cell interactions will facilitate the formation of the extracellular matrix.

3. Assess the morphology and size of 3D spheroids by taking bright field and phase-contrast z-stack micrographs with the Leica DMi8 microscope.

4. Quantify metabolically active cells in 3D spheroids using the ATP luciferase-based CellTiter-Glo® 3D solution over a period of 3 days by measuring bioluminescence with the Glomax luminometer every day (see **Note 6**).

3.4 Salivary Epithelial Differentiation in M3DB

1. Change media to epithelial differentiation medium with FGF-10 to recapitulate SG late development and epithelial morphogenesis.

2. Incubate in epithelial differentiation medium for 8 culture days.

3. Confirm 3D spheroid cell viability using CellTiter-Glo 3D solution as in Subheading 3.3.

4. Assess α-amylase activity in the conditioned media using the EnzChek Ultra Amylase Assay Kit to confirm SG-like secretory function (see **Note 7**).

4   Notes

1. hDPSC lines are commercially available through AllCells (Alameda, CA, USA) and should be used at lower passage to guarantee that their proliferation rate and phenotype are stable.

2. The same lot of FBS should be utilized during the all experiments. Changing the FBS may alter the proliferation of the cell line and the epithelial differentiation process.

3. Retinoic acid was optimized to 5 µM during the first 3 days of culture (3D spheroid formation stage) to achieve an enriched epithelial progenitor population.

4. The concentration of FGF-10 will depend on the FBS lot and on the proliferation rate of the hDPSC cell line. FGF-10 at 40–400 ng/mL can produce an optimal salivary amylase production in vitro.

5. Surface markers for cell proliferation (Ki67), hDPSC, hMSC, and SG-specific progenitors should consistently assessed by flow cytometry through all the different passages.

6. Please follow the manufacturer protocol to optimize concentrations of the CellTiter-Glo 3D solution according to the number of seeded cells.

7. To measure α-amylase secretion upon neurostimulation, please add carbachol (10 µM) to spheroids and then collect conditioned media after 24 h. The activity α-amylase is determined following the manufacturer protocols by measuring the fluorescence after excitation at 495 nm with a microplate reader.

Acknowledgments

The development of this protocol and work was supported by Special Task Force for Activating Research (STAR) for Exocrine Gland Biology and Regeneration Research Group, Ratchadaphiseksomphot Endowment Fund, Chulalongkorn University [grant number: STF 6202432001-1] and Chulalongkorn University Faculty of Dentistry [grant number: DRF62018] in Thailand to J.N.F. and by funding from Mahidol University [grant numbers: E04/2560, 83/2561, and 213/2562] in Thailand to S.R.

References

1. (2014) Chapter 1: Cancer worldwide. In: Stewart BW, Wild CW (eds) World Cancer Report 2014, International Agency for Research on Cancer, France, pp 16–53

2. Mironov V, Visconti RP, Kasyanov V, Forgacs G, Drake CJ, Markwald RR (2009) Organ printing: tissue spheroids as building blocks. Biomaterials 30(12):2164–2174

3. Groll J, Boland T, Blunk T, Burdick JA, Cho DW, Dalton PD et al (2016) Biofabrication: reappraising the definition of an evolving field. Biofabrication 8(1):013001

4. Adine C, Ng KK, Rungarunlert S, Souza GR, Ferreira JM (2018) Engineering innervated secretory epithelial organoids by magnetic three-dimensional bioprinting for stimulating epithelial growth in salivary glands. Biomaterials 180:52–66

5. Ferreira JN, Rungarunlert S, Urkasemsin G, Adine C, Souza GR (2016) Three-dimensional bioprinting nanotechnologies towards clinical application of stem cells and their secretome in salivary gland regeneration. Stem Cells Int 2016 (3):1–9

6. Tseng H, Gage JA, Shen T, Haisler WL, Neeley SK, Shiao S et al (2015) A spheroid toxicity assay using magnetic 3D bioprinting and real-time mobile device-based imaging. Sci Rep 5 (2015):13987

7. Lin H, Dhanani N, Tseng H, Souza GR, Wang G, Cao Y et al (2016) Nanoparticle improved stem cell therapy for erectile dysfunction in a rat model of cavernous nerve injury. J Urol 195(3):788–795

8. Lombaert I, Movahednia MM, Adine C, Ferreira JN (2017) Concise review: salivary gland regeneration: therapeutic approaches from stem cells to tissue organoids. Stem Cells 35 (1):97–105

Chapter 17

Bioprinting 3D Human Induced Pluripotent Stem Cell Constructs for Multilineage Tissue Engineering and Modeling

Jeremy M. Crook and Eva Tomaskovic-Crook

Abstract

Bioprinting human pluripotent stem cells (PSCs) provides an opportunity to produce three-dimensional (3D) cell-laden constructs with the potential to be differentiated in vitro to all tissue types of the human body. Here, we detail a previously published method for 3D printing human induced pluripotent stem cells (iPSCs; also applicable to human embryonic stem cells) within a clinically amenable bioink (also described in Chapter 10) that is cross-linked to a 3D construct. The printed iPSCs continue to have self-replicating and multilineage cell induction potential in situ, and the constructs are robust and amenable to different differentiation protocols for fabricating diverse tissue types, with the potential to be applied for both research- and clinical-product development.

Key words 3D bioprinting, Clinically amenable bioink, Human induced pluripotent stem cells, Multilineage, Tissue

1 Introduction

Bioprinting human pluripotent stem cells (PSCs), whether they are native embryonic stem cells derived from the inner cell mass of a blastocyst [1, 2] or human induced pluripotent stem cells (iPSCs) generated by reprogramming adult cells [3, 4], has significant potential to radically evolve the fabrication of living tissues for research and medicine. PSCs can self-renew and differentiate to all cell types of the body. Furthermore, providing cells with an appropriate microphysiological 3D environment enables emergent properties not attainable with conventional two-dimensional (2D) cell culture. Therefore, combining PSCs with supporting biomaterials to form printable bioinks, together with in situ expansion and differentiation of printed cells to desired lineages within 3D constructs [5], holds tremendous promise for developing a diversity of living tissues to model developmental biology, pathomechanisms,

Jeremy M. Crook (ed.), *3D Bioprinting: Principles and Protocols*, Methods in Molecular Biology, vol. 2140,
https://doi.org/10.1007/978-1-0716-0520-2_17, © Springer Science+Business Media, LLC, part of Springer Nature 2020

and drug efficacy, as well as for replacement therapy following injury and disease.

Here, we describe in detail a protocol adapted from a previous research publication on 3D bioprinting human iPSCs [5]. Our presently explained approach incorporates the same biogel described in Chapter 10 for 3D printing human neural stem cells, which has been adapted to encapsulate iPSCs by similarly cross-linking after printing to form a 3D cell-laden scaffold construct. Importantly, reiterating from Chapter 10, further details of the bioink properties including mechanical and physical properties can be found in our earlier publications [5–7]. Moreover, the present protocol describes methods for in situ iPSC proliferation and characterization. Finally, once prepared the cultured construct can reasonably be differentiated to a tissue of interest using an established method of choice.

2 Materials

2.1 Bioink

1. Alginic acid sodium salt from brown algae (alginate, Al; Sigma-Aldrich) (*see* **Note 1**).

2. Carboxymethyl chitosan (CMC; Santa Cruz Biotechnology).

3. Agarose (Ag; low gelling temperature; Sigma-Aldrich).

4. Phosphate-buffered saline (PBS; pH 7.4; (Sigma-Aldrich)).

5. iPSCs (*see* **Note 2**).

6. Autoclaved 20 mL glass vial and magnetic stir bar.

7. 19 G needle.

8. 1 mL syringe.

9. 15 mL conical tube (Corning).

2.2 Bioprinting

1. 3D computer-assisted design (CAD) software, e.g., Blender open source software.

2. Smoothflow-tapered dispensing tips, e.g., 27 G, 0.2 mm internal diameter (Nordson EFD).

3. 30 cc optimum syringe barrel and piston with tip-end and snap-on end caps (Nordson EFD).

4. Standard 12-well culture plates (Corning).

5. Calcium chloride (Sigma-Aldrich).

6. PBS, pH 7.4 (Sigma-Aldrich).

7. Dulbecco's Modified Eagle Medium (DMEM)/Ham's F-12 Nutrient Mixture (F-12).

2.3 Culture of Printed iPSC-Laden Constructs	1. Printed iPSC-laden construct.
	2. iPSC culture medium, mTeSR1 (STEMCELL Technologies).
	3. ROCK pathway inhibitor, Y-27632 (STEMCELL Technologies).

2.4   Viability Assay

1. Calcein AM (Life Technologies).
2. Propidium iodide (PI, Life Technologies).
3. Tissue culture dish with coverslip bottom (170 μm thickness).
4. Image J (Fiji) software.

2.5   Immunopheno-typing

1. 3.7% (w/v) paraformaldehyde (PFA; Sigma-Aldrich) solution in PBS (pH 7.4).
2. Goat serum (Sigma-Aldrich).
3. Triton X-100 (Sigma-Aldrich).
4. PBS, pH 7.4 (Sigma-Aldrich).
5. Unconjugated primary antibodies:
   (a) OCT4 (mouse, 1:200; STEMCELL Technologies, 60093).
   (b) SSEA4 (mouse, 1:200; STEMCELL Technologies, 60062).
   (c) TRA-1-60 (mouse, 1:200; Millipore, MAB4360).
   (d) TRA-1-81 (mouse, 1:200; Millipore, MAB4381).
6. Goat antimouse IgG (H+L) highly cross-adsorbed secondary antibody, Alexa Fluor 488 (1:1000; Life Technologies).
7. 4,6-Diamidino-2-phenylindole (DAPI; Life Technologies).
8. ProLong Gold antifade reagent (Life Technologies).
9. Single-well/round dishes with coverslip bottom (#1.5, 170 μm thickness) or glass slides fitted with silicone spacer and coverslip (170 μm thickness).
10. Image J (Fiji) software.

2.6   Equipment

1. Magnetic stirrer with heating.
2. CO_2 incubator.
3. 3D-Bioplotter System (EnvisionTEC GmbH) equipped with platform and dispensing head temperature control and contained within sterile biological safety cabinet.
4. Laser Scanning Confocal (Leica TCS SP5 II) equipped with an inverted microscope and Leica Application Suite for Advanced Fluorescence (LAS AF) software.

3 Methods

Preparation of reagents and cell culture should be performed in a biosafety cabinet. Media and reagents should be prewarmed to 37 °C (unless otherwise stated) and maintained sterile. Incubations and culturing should be performed in a 37 °C incubator with a humidified atmosphere of 5% CO_2 in air.

3.1 Bioink Preparation

1. Dissolve 1.5% (w/v) Ag in PBS in a 20-mL glass vial by heating in a microwave oven for 5 s. Allow it to cool and bubbles to disperse. Repeat process with 1–2 s bursts for five to eight times (see **Note 3**).

2. Add 5% (w/v) CMC and stir on a magnetic stirrer with hotplate set at 60 °C for 5 min or until it becomes transparent.

3. Add 5% (w/v) Al and stir at 60 °C for 30 min until it becomes transparent (see **Note 4**).

4. Store the final bioink at 37 °C in dry incubator before introducing cells (see **Note 5**).

5. Transfer an aliquot of 20–40 × 10^6 iPSCs to a sterile Eppendorf tube from previously harvested iPSCs (see **Note 2**).

6. Centrifuge iPSCs at 190 × g for 3 min and remove the supernatant. Gently resuspend cell pellet in 100 μL mTeSR1 medium.

7. Using a 1-mL syringe, carefully add 1 mL bioink solution to a sterile tip-end capped 30 cc syringe barrel.

8. Transfer the resuspended cells to the syringe barrel containing 1 mL bioink.

9. Carefully mix iPSCs by stirring with a 19-G needle to obtain an evenly distributed cell-laden bioink. Once mixing is complete, insert a piston into the syringe barrel and apply a snap-on end cap to seal the syringe.

10. Centrifuge the sealed syringe containing cell-laden bioink at 300 × g for 1 min to remove air bubbles.

11. Cool at 4 °C for 10 min prior to printing.

3.2 Bioprinting and Culture of Printed Constructs

The design of the 3D model should be completed prior to the day of experiment. Before harvesting cells, turn on the EnvisionTEC 3D Bioplotter instrument and set to precool (platform and dispensing head) to 15 °C. Before use, sterilize the Bioplotter instrument work space by UV exposure and wipe down surfaces with 70% ethanol to maintain sterility within biological safety cabinet.

1. Design a 3D model (e.g., 5 mm × 5 mm × 2 mm lattice structure) using CAD software and save the file as STL type.

2. Open the STL file using Bioplotter RP software and convert it to a BPL format file (*see* **Note 6**).

3. Open the BPL file in Visual Machines software to establish new protocol for the 3D printing.

4. Fit bioink syringe containing cell-laden bioink onto a low temperature viscous dispensing head. Remove end-tip cap from syringe and replace with tapered dispensing tip. Follow the manufacturer's instructions to control the 3D-Bioplotter (*see* **Note 7**).

5. Printing parameters must be determined by the end user. As a guide, modify the print speed and dispensing pressure of continuous lines at 90° angle between layers to generate scaffold constructs, having equally sized strands and intervening pores (*see* **Note 7**). A representative printing setup is shown below:
   Pressure = 0.3 bar.

   Preflow = 0.1 s.

   Postflow = 0.1 s.

   Speed = 9 mm.

   Temperature of barrel and platform = 15 °C.

   Pattern = continuous line.

   Distance between strands = 1 mm.

   Layers = 4, rotating 90° for each layer.

   Needle Z offset = 0.25 mm.

6. Once optimal parameters have been determined, extrusion print in air the four-layer constructs of cell-laden bioink into separate wells of a sterile 12-well culture plate.

7. Immediately following printing of each construct, carefully add 2% (w/v) calcium chloride in PBS to cover the printed construct and cross-link the bioink for 10 min at room temperature (RT).

8. Rinse constructs three times with 1 min washes in DMEM/F-12, followed by two successive 10 min washes in DMEM/F-12 at 37 °C in a 5% CO_2-humidified incubator.

9. Finally incubate constructs for 1 h in iPSC culture medium in an incubator, followed by a medium change with fresh iPSC culture medium containing 10 µM ROCK inhibitor, and return to 37 °C in a 5% CO_2 humidified incubator for 3 days culture.

10. Following 3 days culture, replace culture medium with iPSC culture medium and culture for 7 days, with medium changes performed every 2 days (*see* **Note 8**).

3.3 Assessment of Viability

1. Assess iPSC viability for days 1 and 7 after printing by incubating printed constructs with 5 µg/mL calcein AM at 37 °C for 10 min, followed by addition of PI to give a final dilution of 1 µg/mL PI, and incubate for 10 min.

2. Remove labeling solution and wash constructs for 10 min in DMEM/F-12.

3. Transfer the constructs into tissue culture dish with coverslip bottom for confocal microscopy.

4. Calculate the number of live and dead cells using ImageJ.

3.4 Immunophenotyping

To establish pluripotency of iPSC following printing, constructs can be immunophenotyped after a period of iPSC expansion within printed constructs.

1. Aspirate cell medium and rinse constructs with PBS at RT.

2. Fix printed constructs with 3.7% PFA at RT for 30 min and wash twice in PBS.

3. Perform concurrent blocking and permeabilization overnight at 4 °C with 5% (v/v) goat serum in PBS containing 0.3% (v/v) Triton X-100.

4. Prepare primary antibodies at suggested dilution (or as determined by end user) in 5% goat serum in PBS. Incubate separate constructs with each antibody overnight at 4 °C.

5. Wash constructs three times in 0.1% Triton X-100 in PBS.

6. Prepare species-specific Alexa Fluor-conjugated secondary antibody at suggested dilution (or as determined by end user) in 5% goat serum in PBS. Incubate constructs in the dark (wrapped in foil) for 2 h at RT.

7. Wash constructs two times in 0.1% Triton X-100 in PBS at RT.

8. Stain with 10 µg/mL DAPI at RT for 10 min.

9. Transfer the constructs into a tissue culture dish with coverslip bottom for immediate imaging with confocal microscopy. Alternatively, mount constructs onto glass slides fitted with silicone spacer and coverslip applied with ProLong Gold antifade reagent. Image with a confocal microscope.

4 Notes

1. Sodium alginates have a wide range of viscosities, β-D-mannuronic (M) and α-L-guluronic (G) residue ratio, and molecular weight chain distribution that affect subsequent gelation properties. While we have previously used low viscosity alginate with 100–300 cP, M/G ratio 1.67, and a molecular weight

~50,000 Da for optimal iPSC survival during extrusion printing, an optimal % (w/v) ratio of sodium alginate within the bioink must be determined.

2. iPSCs can be harvested from standard 2D cultures using a method of choice. We employ ReLeSR (STEMCELL Technologies) or EDTA (Sigma-Aldrich) to dissociate iPSC colonies, with the latter initially requiring removal of areas of iPSC differentiation from cultures. ReLeSR or EDTA is applied to PBS rinsed cultures at 37 °C in a 5% CO_2 humidified incubator for 5–7 min or 2 min, respectively. Add 1 mL prewarmed DMEM/F-12 (Gibco Life Technologies) to each well. Agitate plates by firmly tapping the side of the culture plate, and transfer cells to a 15 mL conical centrifuge tube. Centrifuge at $300 \times g$ for 5 min. Supernatant is then aspirated and the cells resuspended with fresh prewarmed iPSC culture medium. Determine the number of cells by, for example, trypan blue exclusion assay with a hemocytometer.

3. Keep glass vial covered with lid but ajar to ensure release of gasses during microwave heating. Ensure that agarose solution does not bubble over at any point.

4. The viscous bioink can be additionally stirred with a 19 G needle and vortexed to help disperse any clumps of undissolved material. As needed, centrifuge vial at $300 \times g$ for 1 min to clear remnants of viscous solution from the sides of the vial.

5. The bioink solution can be prepared prior to the day of printing and stored at 4 °C for up to 3 weeks before discarding. Warm bioink to 37 °C in a dry incubator before introducing cells.

6. Create uniform slice thickness of layers equivalent to 80% of the inner nozzle tip diameter.

7. Before printing constructs, it is necessary to calibrate the dispensing tip position, transfer height, and needle offset and to calibrate the needle (dispensing tip) according to the manufacturer's instructions. Needle calibration is required with start-up of machine and after each needle change if required during printing.

8. Following 7 days culture post printing, construct can reasonably be differentiated to a tissue of interest using an established method of choice.

Acknowledgment

The authors wish to acknowledge funding from the Australian Research Council (ARC) Centre of Excellence Scheme (CE140100012).

References

1. Thomson JA, Itskovitz-Eldor J, Shapiro SS, Waknitz MA, Swiergiel JJ, Marshall VS et al (1998) Embryonic stem cell lines derived from human blastocysts. Science 282 (5391):1145–1147

2. Crook JM, Peura T, Kravets L, Bosman A, Buzzard JJ, Horne R et al (2007) The generation of six clinical-grade human embryonic stem cell lines. Cell Stem Cell 1:490–494

3. Takahashi K, Tanabe K, Ohnuki M, Ichisaka T, Tomoda K, Yamanaka S (2007) Induction of pluripotent stem cells from adult human fibroblasts by defined factors. Cell 131(5):861–872

4. Yu J, Vodyanik MA, Smuga-Otto K, Antosiewicz-Bourget J, Frane JL, Tian S et al (2007) Induced pluripotent stem cell lines derived from human somatic cells. Science 318 (5858):1917–1920

5. Gu Q, Tomaskovic-Crook E, Wallace GG, Crook JM (2017) 3D bioprinting human induced pluripotent stem cell constructs for in situ cell proliferation and successive multi-lineage differentiation. Adv Healthc Mater 6:1700175

6. Gu Q, Tomaskovic-Crook E, Lozano R, Chen Y, Kapsa RM, Zhou Q et al (2016) Functional 3D neural mini-tissues from printed gel-based human neural stem cells. Adv Healthc Mater 5:1429–1438

7. Tomaskovic-Crook E, Zhang P, Ahtiainen A, Kaisvuo H, Lee CY, Beirne S et al (2019) Human neural tissues from neural stem cells using conductive biogel and printed polymer microelectrode arrays for 3D electrical stimulation. Adv Healthc Mater 8(15):1900425

Correction to: Characterizing Bioinks for Extrusion Bioprinting: Printability and Rheology

Cathal O'Connell, Junxiang Ren, Leon Pope, Yifan Zhang,
Anushree Mohandas, Romane Blanchard, Serena Duchi,
and Carmine Onofrillo

Correction to:
Chapter 7 in: Jeremy M. Crook (ed.), *3D Bioprinting: Principles and Protocols*,
Methods in Molecular Biology, vol. 2140,
https://doi.org/10.1007/978-1-0716-0520-2_7

In chapter 7, on page 111, an author's name was misspelt as "Yifan Li". It has been updated
as "Yifan Zhang".

The updated original version of this chapter can be found at
https://doi.org/10.1007/978-1-0716-0520-2_7

Jeremy M. Crook (ed.), *3D Bioprinting: Principles and Protocols*, Methods in Molecular Biology, vol. 2140,
https://doi.org/10.1007/978-1-0716-0520-2_18, © Springer Science+Business Media, LLC, part of Springer Nature 2022

INDEX

Printed in the United States
by Baker & Taylor Publisher Services